MATH SMART

MATH SMART

By Marcia Lerner

Random House, Inc.
New York

www.PrincetonReview.com

The Independent Education Consultants Association recognizes The Princeton Review as a valuable resource for high school and college students applying to college and graduate school.

The Princeton Review, Inc.
2315 Broadway
New York, NY 10024
E-mail: booksupport@review.com

 Published in the United States by Random House, Inc., New York, and simultaneously in Canada by Random House of Canada Limited, Toronto.

ISBN 978-0-375-76216-1

Editor: Jeannette Mallozzi
Production Editor: Maria Dente
Production Coordinator: Jennifer Arias

Manufactured in the United States of America.

20 19 18 17 16 15 14 13 12 11

Second Edition

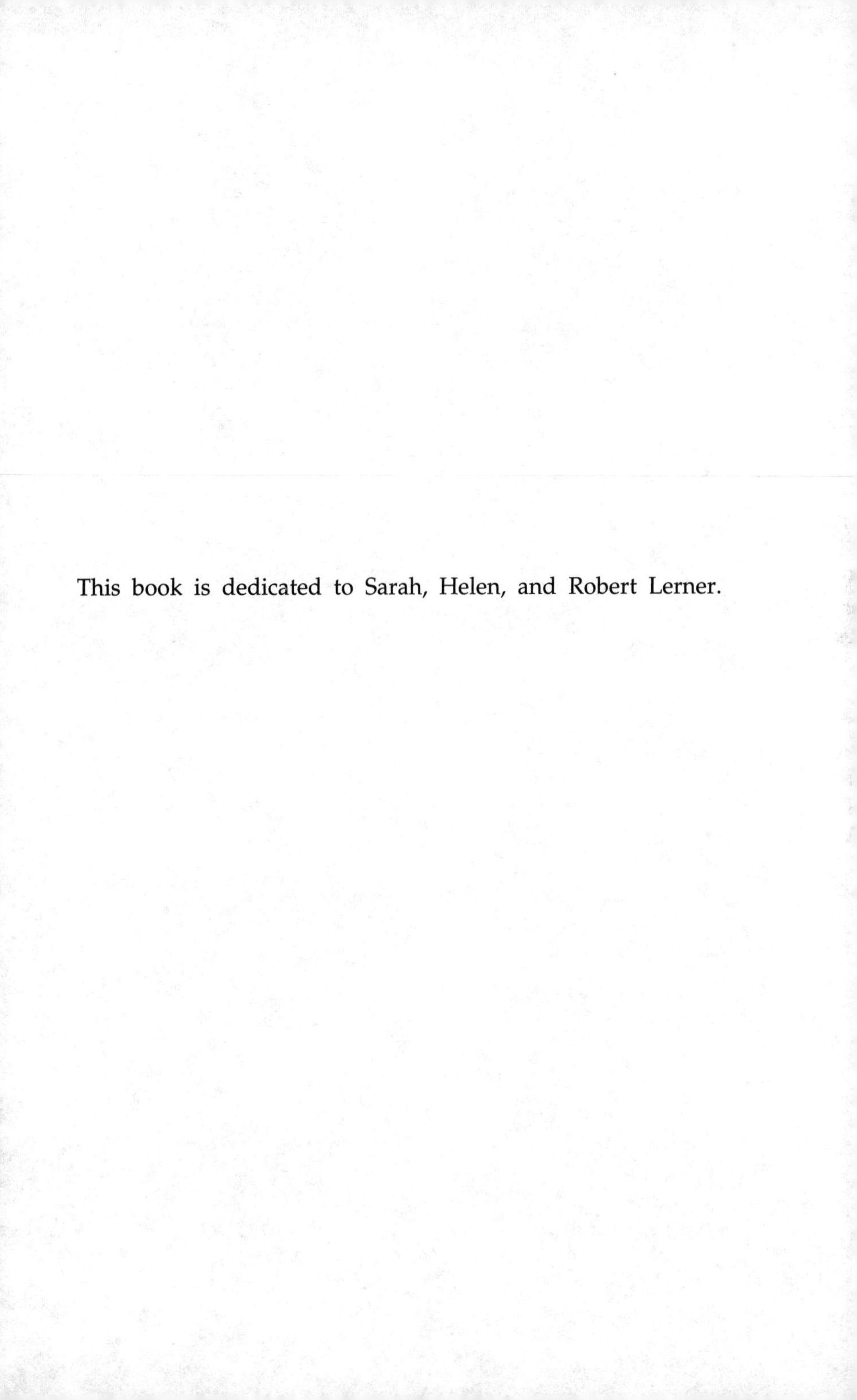

This book is dedicated to Sarah, Helen, and Robert Lerner.

ACKNOWLEDGMENTS

I would like to gratefully acknowledge the major help from and patience of Hanna Fox, Liz Buffa, Nell Goddin, Paul Foglino, and Jessica Kindred. Also Geoff Martz, Julian Ham, Cynthia Brantley, David Bodnick, Leland Elliott, Grace Roegner, Mike Freedman, Jeannette Mallozzi, Maria Dente, Jennifer Arias, Alice Min, and the editors at Random House.

CONTENTS

INTRODUCTION

Do you ever tell people, "Oh, I'm about halfway through the book I'm reading?" Do you ever say, "I don't have enough money to go to the movies?" Can you figure out if you have enough time during a television commercial to go to the bathroom and get something out of the refrigerator? Can you tell which is farther away from you, Massachusetts or California? How about if clothes are too big or too small?

If you have thought or figured out even one of these things, you already understand some things about math.

What does "math smart" mean? Many people equate the ability to do arithmetic quickly with being math smart. But a calculator or an adding machine can perform mechanical functions quickly. Does this mean something with batteries is smart? Unlikely. Being math smart means understanding how math works. This book reviews some basic math functions; it will help you to do these functions and understand why they work. When you're finished with this book, you will have a better grasp of what math means and how to get through it. A lot of people reading this may be saying, "But, the thing is, I'm not really a math person."

People like to think of the world, and their brains, as split into two parts—the math and science part, and the English and history part. It doesn't have to be that way. Being math smart doesn't have to do with some specific kind of brain.

You can read *The Catcher in the Rye* and love it like crazy, and in the next moment figure out that 2 to the third power is equal to 8. Your brain can handle a lot of different things, and just because you are comfortable with one thing doesn't mean you have to be uncomfortable with something else. Math is about patterns that exist in the world—like the veins on leaves having order. When you see patterns in the world, you are understanding math.

You don't have to love math, although if you relax you may find yourself at least liking it more. But it is a sure thing that you <u>can</u> do math.

So sit down, grab some paper, and relax. If you reach something that confuses you, just back up and read the information that comes before it. Or take a walk around the block, listen to some music, read something else for a while. You can always come back.

How to Use This Book

This book covers the basic underpinnings of math. It covers all the basic math found on the standardized tests you may face trying to go to college (SAT), graduate school (GRE), or business school (GMAT). There are drills throughout, and at the end of each chapter there are review questions for you to practice on. If there is a chapter that covers something you already feel comfortable with, just do the review questions to make sure, and then go on to the next chapter. You will also find answers and explanations (when necessary). At the end of every chapter there is also glossary of the terms covered, along with the page numbers noting where the term was introduced. There is also an index in the back of the book.

Why the Princeton Review?

We are the world's leader in test preparation. Each year we help more than two million students score higher through our courses, award winning software and bestselling books. Our philosophy is simple: Teach students exactly what they need to know while providing them with an experience that is truly unique and FUN. And that's exactly what we plan to do.

Enjoy!

CHAPTER 1

NUMBERS

First get yourself some nice clean supplies, such as a fresh pad of paper, some sharp pencils, and one of those giant pink erasers. Not that you need this stuff specifically, it's just that nice supplies tend to make you feel better when you do things. Just make sure that whatever kind of math you do, you **write everything down**. Writing down math and making it visual are some ways to make math something you can do easily, instead of something that makes you sweat.

You probably don't realize how much you use math in everyday life. You use math when you decide how much time you have to get somewhere, when you decide if you have enough money for both a slice of pizza and a soda, or when you figure out how many minutes need to pass before you get out of class.

Consider this:

> Jenny says, "Hey, Susan, let's take the bus to the record store downtown and get the new Beck CD."
>
> Susan says, "No way, I can't do that. I only have ten dollars and the bus is two dollars and the CD is eight something with tax."
>
> Jenny says, "Oh that's cool, I have twenty dollars, I'll cover you."

Get this: Jenny is getting a D in her math class. She may not know it, but she just demonstrated that she has the capacity to do way, way better. The thing is, Jenny understands math really well—she just doesn't realize that the skills she uses in her normal life are math. Planning, guessing, and approximating—these are all skills she could bring to math class.

Here's what Jenny did when she made plans with Susan:

$$\left(2\left[2+8+(8\times .0825)\right]\right)=21.32 \quad \text{and} \quad 21.32<30$$

Jenny and Susan's plans may look sort of scary when laid out in number form like that, but remember, **if you can figure out whether you have enough money to cover your friend, you can do math**.

The only thing that makes math different from what you already do is that math sometimes uses numbers without telling you they relate to a CD, or a bus fare. Math uses numbers without references. But using numbers solo can actually make things simpler. Here's how.

NUMBERS

The numbers you will use in these first few chapters are **integers**.

Numbers like -1, -400, $-10,000$ are integers. 0 is an integer.

But $\frac{1}{2}$, -0.089, and 5.78 are not integers, these are fractions and decimals.

An integer is any number that is not a fraction or a decimal.

Positive numbers, numbers larger than 0, can be integers. Numbers like 73, 1, and 5 are integers.

Negative numbers, which are numbers smaller than 0, can be integers.

Whole numbers are all the positive integers and 0.

The easiest way to visualize integers is to put them on a number line. You probably remember these. Here is a number line:

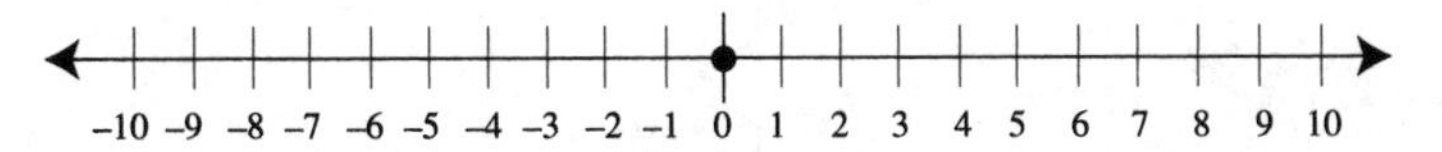

Moving to the right the numbers become larger, moving to the left the numbers become smaller.

Here is the number line that will be used in this book:

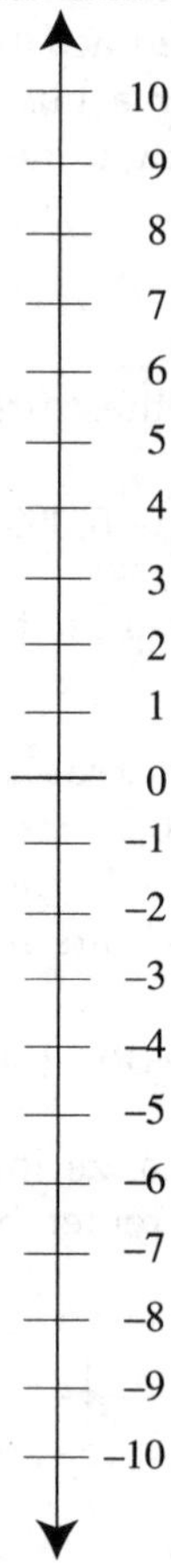

The ends of a number line go on forever, because there is no end to the numbers in the world. The way to express this is to write the symbol ∞, which represents infinity. (Don't think about this too much or it will blow your mind.) By making the number line vertical, it's easy to see which numbers are larger. Just like using a measuring stick, any number that's higher up is bigger.

No matter where you are on the number line, any number *above* another *is* the bigger of the two. This is always true, no matter whether the numbers are positive, (bigger than 0), negative, (less than 0), or even 0.

Why is –1 bigger than –10? If you owe someone $1, you have more money for yourself than if you owe someone $10. Owing someone money is a good way to think about negative numbers. Look on your number line and see which number is higher. This will help when you have any kind of confusion over which of two numbers is bigger. Think about –100 and –1,000. Put them both on your own number line. You don't have to write out a million-mile-long number line for this because you can change the scale.

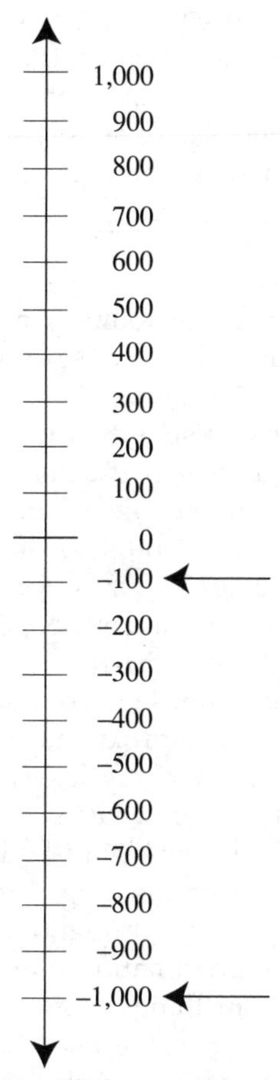

Which number is higher? –100. So this is the larger number.

The distance on the number line from 0 to the –100 is 100 spaces. This is also known as the **absolute value**. Even though these 100 spaces are on the negative part of the line, you refer to them as the absolute value of -100, which is positive 100. The absolute value of any number is always positive, whether the number itself is positive or negative. The way to write the absolute value of -100 is this:

$$|-100| = 100$$

The absolute value of 7 is $|7| = 7$.

You might notice that when negative numbers are written, they have the negative sign next to them, like -5, but when positive numbers are written, most won't have the positive sign; it will be just the number. That's simply the way these numbers get written, so don't worry about it.

THINK ABOUT THIS

One of the most useful tools in math is one that everybody uses and almost nobody really considers math. It's **approximation**.

Approximation is guessing the rough size of a number instead of knowing it exactly. When you approximate how many people are in your house you may not know the exact number of people, but you know it's either 2 or 3. Or maybe you're having a party and there are around 50 people in your house. Or you're having a major party and there are around 500 people in your house. Your math skills will improve tons if you approximate even more than you already do. How many leaves are there on a particular tree? How many beans do you get served with dinner? How many hard-boiled eggs would fill the room you are sitting in right now? (We know, we know, that's gross.) Throughout this book there are going to be approximation questions regarding all kinds of different things. Give them a try, and then find the answer in the back to see how close you get. And how many people *are* in your house right now, anyway? Approximately.

DIGITS

Digits are the integers 0 through 9. The places in a number are called **digit places**. For instance, 357 is a three-digit number, because it takes up three digit places; 200 is also a three-digit number, though you need only two digits, 2 and 0, to write it. It is a three-digit number because it takes up three digit places. Each digit place has a name. In the number 357, 7 is what you call the **units digit**. You can also call it the **ones digit**. The **tens digit** is 5, and 3 is the **hundreds digit**.

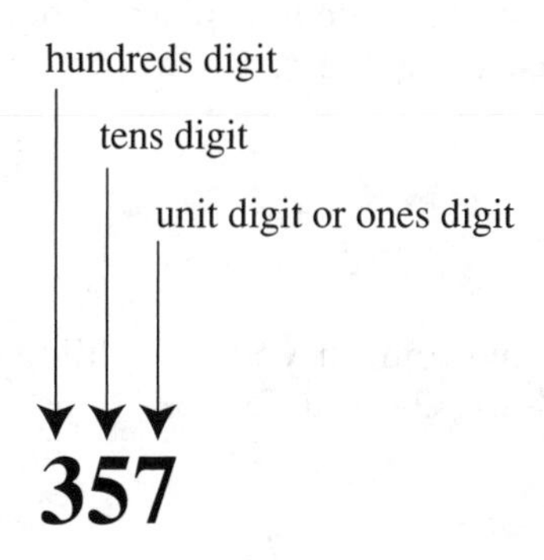

What is the units digit of 456?
What is the tens digit of 300?
What is the hundreds digit of 701?
You can find the answers in the answer key at the end of the chapter.

CALCULATIONS WITH INTEGERS

If you understand what we were just talking about this next section will be no problem.

ADDITION

When you **add** positive integers, you combine them to form a larger number. Addition is putting integers together to get their **sum**. Sum is the result of an addition. Think of addition with Frisbees and it will be easy. If you have 5 Frisbees and someone gives you 3 more, you end up with 8 Frisbees. You can use Frisbees, apples, or hamburgers. Just put the things together and count how many there are.

$$\begin{array}{r} 5 \\ +\,3 \\ \hline 8 \end{array}$$

When you have a stack of numbers you are combining, you add the units digits. When you add bigger numbers, you sometimes have to **carry over**. For instance, you used to have 7 shirts. Then, for your birthday, you got 5 more. After your birthday, how many do you have? 12. When you add the two units digits together, you take the 1, or the tens digit from this number, and put it in the tens place. That is what carrying over means, transferring the tens digit to the tens place.

$$\begin{array}{r} 7 \\ +\,5 \\ \hline 12 \end{array}$$

You do the same thing if you are adding numbers greater than 10. Try adding 24 and 78.

$$\begin{array}{r} 24 \\ +\,78 \\ \hline \end{array}$$

You add the units digits, 4 and 8, and get 12. The tens digit of 12 is 1, so you add it to the tens digit of the numbers you are already adding.

$$\begin{array}{r} {}^{1}24 \\ +\,78 \\ \hline 2 \end{array}$$

You have 1 along with the 2 and the 7 that are already there, and they add up to 10. The 0 becomes the tens digit, and the 1 becomes the hundreds digit.

$$\begin{array}{r} {}^{1}24 \\ +\,78 \\ \hline 102 \end{array}$$

Adding more than two numbers works the same way. Try adding

$$\begin{array}{r} 14 \\ 32 \\ 54 \\ +\ 6 \\ \hline \end{array}$$

Add the numbers in the units digit column and get 16. Put the 6 as the units digit of your new number and carry the 1 from 16 over to the top of the tens column.

$$\begin{array}{r} \overset{1}{1}4 \\ 32 \\ 54 \\ +\ \ 6 \\ \hline 6 \end{array}$$

You add that 1 to the 1, 3, and 5 that are already in the tens column and you get 10. You put the 0 in the tens place, and the 1 in the hundreds place. You have 106.

$$\begin{array}{r} \overset{1}{1}4 \\ 32 \\ 54 \\ +\ \ 6 \\ \hline 106 \end{array}$$

When you are adding stacks of numbers, always make sure that your digits places are lined up so when you carry digits, they go to the right place.

APPROXIMATE THIS:

How many chocolate bars would you have if you started with 3,524 and someone gave you 99,999?

Remember, answers to all "approximate this" questions are in the back of the book.

SOMETHING TO THINK ABOUT

Take a minute to think about how carrying digits works.

When you get 16 from adding the numbers in the ones column, from 4, 2, 4, and 6, why do you take the 1 from 16 and add it as a 1, instead of a 10, from 16?

Well, since you are adding it to the tens place, it is like you are adding 10. A 1 digit in the tens place means 10. Take a look.

10.

The only number other than 0 in 10 is 1, but that doesn't mean the value of that number is 1. When the digit 1 is in the tens place, and there is a 0 in the units place, the number is 10. If the 1 were in the hundreds place, what would the number be then?

100.

This isn't what most people think about when they are "carrying the 1" or "carrying the 2" over, but the more you understand why math works the way it does, the easier things will be for you, and the sooner you will teach yourself to stop panicking whenever you come across anything vaguely mathematical.

When you see an addition problem written out like a sentence, make it vertical so that your digit places are lined up right. In fact, when you see a problem written out, think of what the amounts are and approximate the answer. Then line your numbers up on top of each other, add them, and see how close your approximation was. Try a few addition problems, until you are sure you feel comfortable.

Quiz #1

1. $123 + 345 =$
2. $324 + 777 =$
3. $678 + 5 =$
4. $134 + 45 + 12 + 9 =$
5. $499 + 3 =$
6. $100 + 10 + 1 =$
7. $14 + 7 + 232 + 1 + 132 =$
8. $7,000 + 6,999 =$

More Things to Think About

Sometimes you can make adding really big numbers easier. For instance, let's try with medium-sized numbers, 15 and 17. To do this quickly in your head, first try to make it add up to an easy number. For instance, 15 is an easy number to work with, but 17 is a harder number.

Remember, numbers can be separated into smaller pieces just like they can be combined to form bigger pieces: 17 is just $10 + 7$, or $5 + 12$, or $8 + 9$. Thinking about numbers this way often makes life easier when you need to deal with really huge numbers, or to think about approximating.

You have 15. You have 17. The 17 is made up of $15 + 2$, so take the 15 first and add it to the other 15.

$$15 + 15 =$$

It gives you 30, and, you still have the 2 from the 17 left. It's a lot easier, though, to add 2 to 30 than it is to add 15 to 17.

Separating these numbers also brings you to the next type of calculation . . . .

SUBTRACTION

The next thing you want to do without fear is take numbers away from each other, or **subtract**. Subtraction is the **inverse** (reverse) operation of addition. If you have 7 pieces of gum, and your friend took 2 of them, you would be left with 5. That's subtraction. Look at it on the number line.

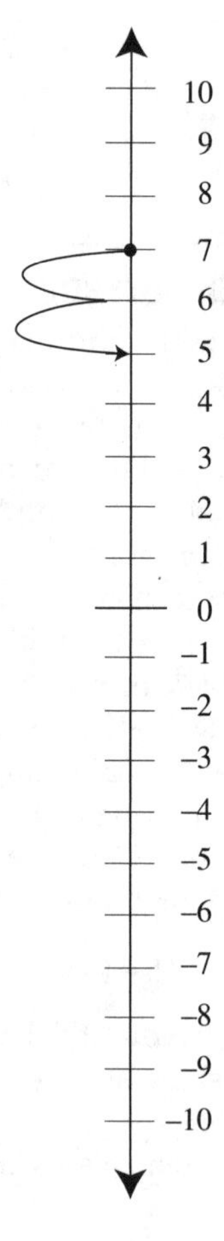

On the number line to subtract or take away a positive number, you move down.

And setting up subtraction problems is just like addition; you line up the numbers along the units digit.

$$\begin{array}{r} 7 \\ -\,2 \\ \hline 5 \end{array}$$

When you are dealing with numbers whose units digits don't subtract well, it is a little different.

$$\begin{array}{r} 20 \\ -\;\;7 \\ \hline 13 \end{array}$$

You are not subtracting 7 from 0, you are subtracting 7 from 20. When you have 20 and subtract 7 you are left with 13. You can go grab a group of 20 cookies or something, and take away 7, and see. The way you make this work in a subtraction problem is to **borrow**. You borrow whenever the digit you are subtracting is bigger than the digit you are subtracting from. In this problem, you borrow 1 from the 2 which is in the tens place, to make the units digit of 0, bigger. You leave the leftover 1 in the tens column. Together the two 1s make up the 2 that is in the tens place. Remember the 2 in the tens place really means 20.

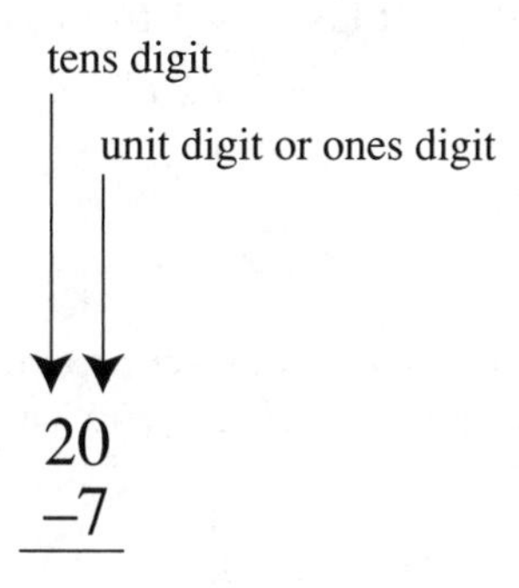

You now have 10 minus 7, which gives you 3. You also have a 1 still left in the tens place because you started out with 2.

$$\begin{array}{r} {\scriptstyle 1\,1} \\ \not{2}0 \\ -\;\;7 \\ \hline \end{array}$$

Since there is no digit in the tens place of 7, you are subtracting nothing from the 1. You are left with that same 1 in the tens place of your answer, which gives you 13.

$$\begin{array}{r} \overset{1}{\not{2}}\overset{1}{0} \\ -\ \ 7 \\ \hline 13 \end{array}$$

You can borrow just the same when there is a 0 in the tens or the hundreds place. Here's how it looks:

$$\begin{array}{r} 100 \\ -\ \ \ 7 \\ \hline \end{array}$$

To subtract the 7 from the 0, instead of borrowing a 1 from yet another annoying 0, you treat the hundreds and the tens digit as one number, the number 10.

$$\begin{array}{r} \not{1}\overset{9}{\not{0}}\overset{1}{0} \\ -\ \ \ 7 \\ \hline \end{array}$$

Grab a 1 for your units place 0 and get $10 - 7 = 3$, and what do you have left after you take the 1 from the 10? 9, definitely.

$$\begin{array}{r} \not{1}\overset{9}{\not{0}}\overset{1}{0} \\ -\ \ \ 7 \\ \hline 93 \end{array}$$

So, $100 - 7 = 93$. If you are subtracting a number that involves carrying from more than one place, you will still be fine. Like:

$$\begin{array}{r} 1{,}030 \\ -\ \ \ 35 \\ \hline \end{array}$$

Borrow a 1 from the 3 in the tens place for the units place 0, and you get 10 minus 5, which is 5.

$$\begin{array}{r} {\scriptstyle 9\;12\;1} \\ 1{,}030 \\ -\quad 35 \\ \hline \end{array}$$

Now you have a 2 in the tens place, because you borrowed a 1, and to subtract 3 from 2 you need to borrow another 1 from the 10 in the hundreds and the thousands place. You have 12.

$$\begin{array}{r} {\scriptstyle 9\;12\;1} \\ 1{,}030 \\ -\quad 35 \\ \hline \end{array}$$

12 minus 3 is 9, and you still have the 9 left over from when you borrowed 1 from the 10 in the hundreds and thousands place.

$$\begin{array}{r} {\scriptstyle 9\;12\;1} \\ 1{,}030 \\ -\quad 35 \\ \hline 995 \end{array}$$

For a quicker way to do this see More Things to Think About on page 16.

You can subtract using more than two numbers, like 100 – 50 – 25—but, beware! You can't stack numbers and then subtract them, the way you do with addition. Instead when you are subtracting a long list of numbers, go from the first subtraction to the last, left to right, only two numbers at a time.

$$\begin{array}{r} 100 \\ -\ 50 \\ \hline 50 \end{array}$$

$$\begin{array}{r} 50 \\ -\ 25 \\ \hline 25 \end{array}$$

Try subtracting some more:

Quiz #2

1. $158 - 92 =$
2. $34 - 4 =$
3. $123 - 67 =$
4. $23 - 22 =$
5. $4,567 - 3,759 =$
6. $725 - 627 =$
7. $35 - 16 =$
8. $2,040 - 45 =$

More Things to Think About

Look at that last question, $2,040 - 45$. You can use a shortcut here, just the same as you can in an addition question. Instead of thinking about it as 2,040 minus 45, think of an easier number to subtract from 2,040. How about 40?

$$2,040 - 40$$

It leaves you with an even 2,000. Now, what you have left of your 45 after subtracting the 40 from it is 5.

$$2,000 - 5$$

You have 2,000 minus 5. Remember, you can look at numbers in many different ways, and this is one thing you should think about if you want math to become a less scary part of your life.

APPROXIMATE THIS:

How much does the earth weigh?

CALCULATING ALL THIS STUFF WITH NEGATIVE NUMBERS

Okay, you're thinking "Fine fine fine, I have no problem with all this but what about the really horrifying part, doing all this with NEGATIVE numbers, or with GIGANTIC numbers, or . . ."

Relax. If you can do all the calculations so far, you will have no trouble with the rest of the calculations in this book, or in all the rest of math, really, because these are the foundations.

ADDING NEGATIVE NUMBERS

With whole numbers you can say, "Well 3 is like having 3 baseballs or 3 bananas or whatever." But what is a good way to visualize –3?

Think of –3, or any negative number, as the number owed. For instance, if you have no money in your pockets, AND you owe your best friend $3, it is like having –$3. And that's how you can think about adding them. Say you had –$3, and someone came along and gave you $5. Since $3 of those $5 would be absorbed by the money you owe your friend—if you're being prompt and honest, and for the problem's sake we'll assume you are—then all you end up with is $2 for yourself.

$$\begin{array}{r} 5 \\ +\ (-3) \\ \hline 2 \end{array}$$

The easiest way to look at adding a negative number to a positive number is that you are letting the negative sign take over the addition sign, which makes it a subtraction problem.

$$\begin{array}{r} 5 \\ -\ 3 \\ \hline 2 \end{array}$$

The reason this works takes you back to the number line. When you are adding positive numbers, you are taking a number that is above 0, and moving it even farther up on the number line. In fact, if you have 3 and you add 4, you are moving from 3 on the number line, four spaces up, to end at 7.

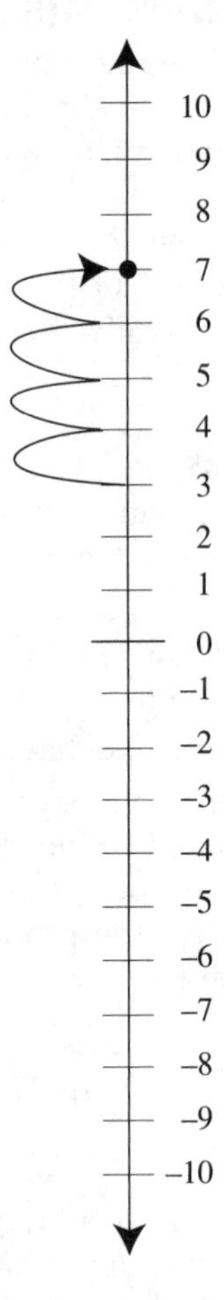

This is because 3 is three spaces above 0 on the number line, and 4 is four spaces above 0 on the number line, so if you combine them . . .

BUT, if you have 3 on the number line, and you are adding –2, which is two spaces below 0 on the number line, you have to move two spaces down (or fewer) from 3 when you combine them. In this way, it is just like subtraction, moving down instead of up.

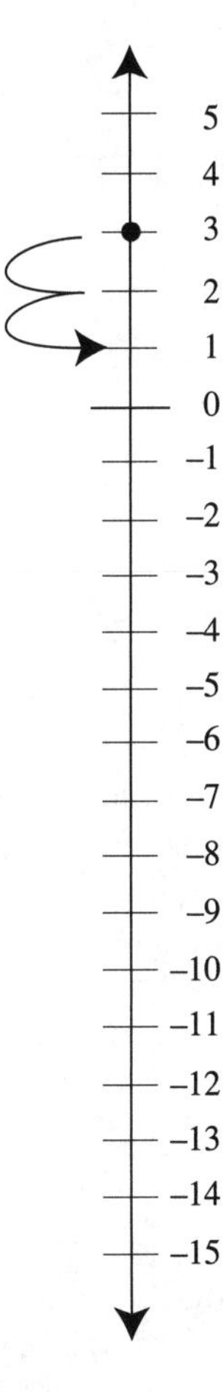

What happens when you add two negative numbers? Try it out on the number line and see. Let's say you add –5 and –7.

$$\begin{array}{r} -5 \\ + \; -7 \\ \hline \end{array}$$

Find –5 on the number line. Then, since you are adding –7, move seven places down. Where do you end up? –12.

If all the numbers in your problem are negative, you can add them like they are positive numbers. Just remember, keep the negative sign next to your sum.

Try adding these numbers:

Quiz #3

1. $5 + -6 =$
2. $-6 + -8 =$
3. $-5 + 4 =$
4. $44 + -33 =$
5. $-12 + -3 =$
6. $13 + -7 =$
7. $-27 + -14 =$
8. $3 + -31 =$

It makes sense to use the number line as often as you need to. Using the number line and writing down the math as you work can make math clearer. Finding a way of thinking about math that is visual, or tangible—using a number line, money, Frisbees or hamburgers, or whatever works for you—can make it easier to understand. Try any way of thinking to make math clearer! The best way to get comfortable with math is to realize that math is simply a way of looking at all the things that are part of life around you.

APPROXIMATE THIS:

Jackie owes her parents $1 million (1,000,000) because she blew out the speakers on their incredibly expensive stereo, and she gives them $1,000 a year (they are not charging her interest). How long would it take her to pay them back?

SUBTRACTING NEGATIVE NUMBERS

Think back to that situation where you owe your friend $3. And let's say, you still haven't paid her back, and your brother comes along and reminds you that you also owe him $5. You now have a grand total of –$8. "But don't worry," he says, "since you painted my house for me last week I am going to wipe out that debt." That means you have $-\$8-(-\$5)$. You are back up to where you started. You owe only \$3.

$$\begin{array}{r} -8 \\ \underline{-(-5)} \\ -3 \end{array}$$

Try it out on the number line. Find −8, and then move five places up. Why up? Think about how you move when you subtract.

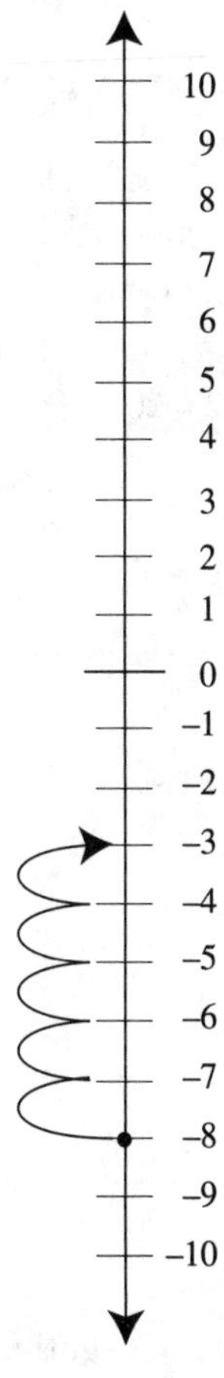

If you had 9 and you were to subtract 7, you would move 7 down. It's as though you are erasing seven of the places that are positive numbers.

It is the same when you subtract negative numbers, except to get rid of negative numbers you have to move up the number line. What this means is that one way to look at subtracting a negative number is to think of it as adding a positive number. Your number is getting bigger, the more negative numbers you subtract. Think of it as getting rid of debt.

$-8-(-5)$ is the same as $-8+5$. The parentheses are used so the problem looks pretty and doesn't have two subtraction signs next to each other.

Subtraction works the same way when subtracting negative numbers from positive numbers.

Let's say you have $20 in cash, but you owe your Aunt Helen $2. That would be $20+(-2)=18$. Now your aunt walks by and says you can forget about the debt, she is going to erase it because you made her a chocolate cake: $18-(-2)=20$. It's the same as saying $18+2=20$.

Try a few of these right now:

Quiz #4

1. $-5-(-4)=$
2. $-14-(-15)=$
3. $-1-(-10)=$
4. $-7-(-3)=$
5. $2-(-5)=$
6. $13-(-1)=$
7. $23-(-23)=$
8. $15-(-17)=$

What if you start with a negative number and subtract a positive number?

Try −37 − 12.

You already know how to subtract positive numbers. Move down the number line, and your original number gets smaller.

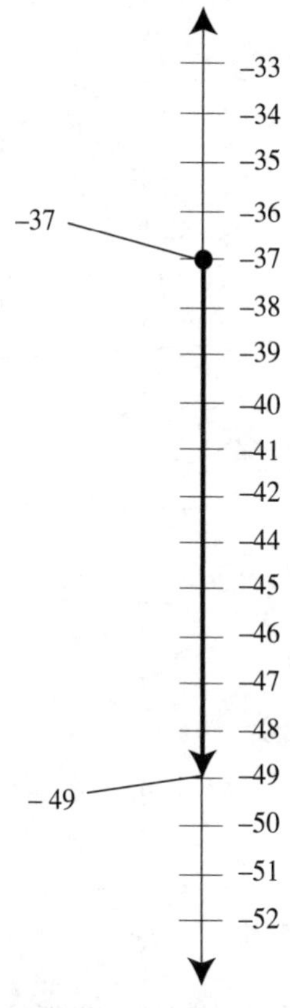

Go to −37 on the number line and move twelve spaces down. You get −49.

An easy way to look at these problems is to add the numbers as though the negative and the subtraction signs weren't there. Then stick on the negative sign.

You can do a shortcut like this when you are subtracting positive numbers from each other, too. You will need it when you are subtracting a larger number from a smaller number. What happens when you have 4 hamburgers and you subtract 6 hamburgers? You end up with a negative number of hamburgers. Try it on the number line.

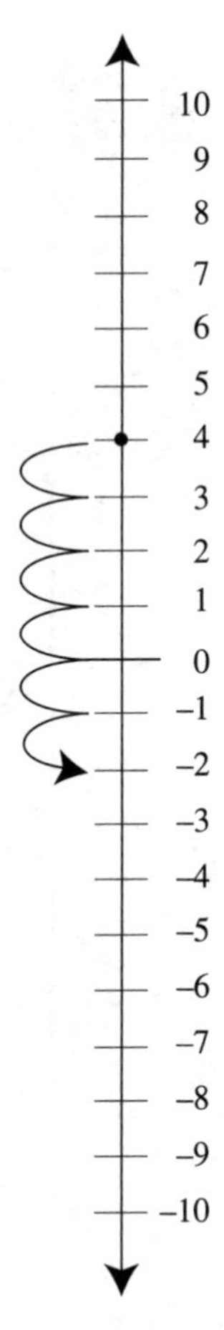

You start at 4 and move six places *down*, because you are subtracting. You end up at –2.

THINK OF THIS

The shortcut to this calculation is to rearrange the numbers in your own way. Put the bigger number, in this case 6, in front, and put the smaller number, in this case 4, after it.

$$6 - 4 =$$

Now subtract them as you normally do. You get 2. BUT, since you remember that you were originally subtracting bigger from smaller, you stick on a negative sign and get -2. Neat, huh? The reason that this works is that subtraction is finding the difference between two numbers. Whether this difference is positive or negative depends on the numbers in the particular equation.
The important thing is to keep track of the sign of the number, whether it is negative or positive, and keep track of the sign of the calculation, whether it is addition or subtraction.

Try subtracting these numbers:

Quiz #5

1. $5-(-6)=$
2. $-12-(-17)=$
3. $-30-5=$
4. $68-74=$
5. $5-10=$
6. $-13-5=$
7. $32-(-15)=$
8. $-3-(-13)=$

Now you can look at all these calculations together. You need not fear dealing with a lot of those irritating long arithmetic problems, as long as you write everything down in a clear way.

Draw out your number line when you are unsure whether a number gets bigger or smaller, and look at the problems piece by piece instead of all at once. You will be absolutely fine.

Try it with this problem.

$$43 - 23 + 45 - 3 + 7 =$$

The problem looks like it might get complicated, but just move one piece at a time.

$$43 - 23 = \_\_$$

What do you get? Now take this new number, the difference between 43 and 23, and use it in the next step.

$$\_\_ + 45 = \_\_$$

Your number plus 45 equals what? Now, use this sum in your next step.

$$\_\_ - 3 = \_\_$$

You've only got one more step to go. Grab that difference you got between your last number and 3, and put it together in this last operation.

$$\_\_ + 7 = \_\_$$

And there you are with your answer. Did you get 69? If you did, excellent job! If you are a little bit off, try going back to each step so you can figure out where your calculation went off.

Did $43 - 23 = 20$? Did $20 + 45 = 65$? Did $65 - 3 = 62$? Did $62 + 7 = 69$?

Once you find your error, give it another try. Being careful and writing things down might seem too boring to even deal with, or too annoying and time consuming. But these are just tools, just like sprints is boring, too, but the only way to play great basketball is to put all the smaller pieces together. The more you get used to calculations, and the more you feel comfortable with the whole idea of numbers—and don't worry, you will—the faster and easier math will become.

Try doing the same step-by-step calculations with the next few problems and take your time! Use tons of scratch paper to write down every tiny thing.

Quiz #6

1. $43+5+5+1-3-6+2=$
2. $(-2)-(-4)+5-6+7=$
3. $2-33+(-4)+5-(-6)-70=$
4. $3+5-2+4=$
5. $5-7-8-9-10=$
6. $107-103-5+4-(-4)+17=$
7. $3+4-4+4-4=$
8. $22-(-7)+5-(-12)+5=$

Here is a load of problems. These questions combine everything you have just read about. A lot of people get nervous when they see what they call "word problems," but remember, word problems are the way everything so far has been explained. You owe your friend $5 . . . You have 6 hamburgers . . . You have to buy a CD and you have $10 . . . It's all the same stuff. To prove it, after these problems are done, you can check the answer key where everything is translated into addition and subtraction. Give these a try:

Quiz #7

1. John has 3 hamburgers and gives away 2 to Cheryl. He then gets 3 more from Bob, and 5 more on top of that, from Gregory. He decides to give 3 to his brother Bill, how many hamburgers does he have left at the end?
2. Barbara starts out the day with 15 hats. She gives away 5, gets 12 in a new shipment, throws out 7, and gets 3 more in a sidewalk sale at the end of the day. How many hats does she have at the end of the day?
3. Sam has no money, AND, he owes his boss $25. Julie gives Sam $50 for his birthday. He pays back his boss and then buys himself a beer for $6, (it's imported from far, far away). How much money does he have now?

4. You had 20 minutes before your class starts and you used 7 minutes to talk to your friend, 2 minutes to buy a candy bar, and 11 minutes to space out and stare out the window. How many minutes are left before class?
5. Twelve dishes are in the sink from last night, and you are supposed to wash them all. If you make your friend Mike wash 2, you throw out 3 that have disgusting stuff caked on them, and you wash 4 yourself, how many dishes will be left in the sink?
6. Patricia has 3,241 jelly beans in her room in the drawer of her desk. If she eats 3 tonight, 20 tomorrow, 53 the next day, and 71 the day after that, and then she swears never to eat another jelly bean and locks all the rest up in a big safe, how many will she end up locking up in that safe? (Yes, it does seem like Patricia might have a little bit of a problem.)
7. Jennifer owes her friend Alice $500. She gets her paycheck of $275, and pays it all to Alice. Then she gets $35 from her parents and gives Alice that. Then she gets $117 in prize money from a raffle and gives Alice that. Does she still owe Alice money, and if she does, how much?
8. George has 3 hours—180 minutes—before he has to get to work. He spends 25 minutes at the gym, 30 minutes at his friend's house, 80 minutes at a disreputable pool hall, and 50 minutes eating lunch. Will he make it to work on time?

It went fine, right? It is all addition and subtraction. Don't worry about people who can do these kinds of calculations in fifteen seconds; speed of calculation has nothing to do with intelligence, it's just a talent that some people have.

Understanding math has to do with figuring out how the numbers work together, not with doing anything really fast.

MULTIPLICATION

You may not have looked at it this way before, but **multiplication** is actually just an outgrowth of addition. It's a way of saving time. The result of multiplication is called the **product**.

Look at it this way: if you were to add 5 and 5 and 5 and 5 you would end up with 20.

$$5+5+5+5=20$$

But that takes a long time, particularly if you wanted to add twenty 5s together, or a hundred 5s. That's where multiplication comes in. Multiplication can shorten the work of adding four 5s together. Just like in a grocery store when you get four cans of soup for the same price, they say "Soup, four times." You can say "5, four times" or 4 times 5 or $4 \bullet 5$ or 4×5 or even $(4)(5)$. Using a big dot or an $\times$ or the numbers in parentheses right next to each other are all different ways of saying "multiply by."

$$4 \times 5 = 20$$

The product of 4 times 5 is 20.

Now that you know what the pieces of multiplication mean, how do you multiply any old number? The easiest way to handle multiplication is to memorize the times tables. We know you probably had to memorize them in third grade and now they seem idiotic and dull. It's true, memorizing is a drag, but once you have the tables through twelve memorized, you will find the rest of multiplication really, really easy. And don't worry about "Well is this really the way math smart people deal with multiplication?" It is. This way, you the reader, who are fast becoming among the math smart in the world, don't need to worry about the annoying mechanics of problems. You can just concentrate on understanding them.

A Couple of Things to Think About

Any number multiplied by 1 is itself. It makes sense, 5, one time, is 5.

Any number multiplied by 0 is 0. If you had 5 zero times, how many would you have?

	2	3	4	5	6	7	8	9	10	11	12
2	4	6	8	10	12	14	16	18	20	22	24
3	6	9	12	15	18	21	24	27	30	33	36
4	8	12	16	20	24	28	32	36	40	44	48
5	10	15	20	25	30	35	40	45	50	55	60
6	12	18	24	30	36	42	48	54	60	66	72
7	14	21	28	35	42	49	56	63	70	77	84
8	16	24	32	40	48	56	64	72	80	88	96
9	18	27	36	45	54	63	72	81	90	99	108
10	20	30	40	50	60	70	80	90	100	110	120
11	22	33	44	55	66	77	88	99	110	121	132
12	24	36	48	60	72	84	96	108	120	132	144

Yikes, we know. But there are a couple of great things. For instance notice the units digits in the 9s column. Notice anything fishy? How about comparing 9 times 2 to 9 times 9. And 9 times 3 to 9 times 8? Interested in reversals much? And look at any number times 2, and that same number times 12. They both end with the same digit, right? There are all kinds of great patterns you can find in multiplication tables, and patterns, as you get used to them, will make math and all things associated with math easier and easier.

Anyway, once you have the whole multiplication table at your fingertips (you might want to carry flash cards around till you get them all), you can start multiplying bigger and bigger numbers. Numbers like 7,639,243,662,172. But to handle the big numbers' league, let's start with 54.

Now say you had 54 relatives, and the holidays were coming up, and you had to buy each and every one of them a present. Now say you didn't have a whole lot of money and you decided to get each of them something lame, like a$3 washcloth. To figure out if it's even worth doing this, you want to figure out how much it is going to cost you. So what, as an officially math smart person, do you do? You multiply: 54 people multiplied by $3. Here you can use something called the **distributive law**. This law says that multiplying 54 times 3 is the same as multiplying 4 times 3 and adding it to 50 times 3. You are separating the 54 into two easier pieces.

$$\begin{array}{r} 4 \\ \times\ 3 \\ \hline 12 \end{array} \qquad \begin{array}{r} 50 \\ \times\ 3 \\ \hline 150 \end{array} \qquad \begin{array}{r} 150 \\ +\ 12 \\ \hline 162 \end{array}$$

4 times 3 is 12.
50 times 3 is 150.
54 times 3 equals 162.

Excellent job, but probably not worth it for the washcloths, right? Now let's get back to that distributive law. What does the distributive law mean, really?

Officially, the distributive law says that: $a(b+c)=ab+ac$.

What this means is that 5×12 is equal to 5 times 12 broken down in the easiest way you know. Like $7+5$, or $6+6$, or $10+2$.

$$5(10+2)=$$

$$(5\times10)+(5\times2)$$

or

$$50+10$$

Together these equal 60. Just like $5 \times 12 = 60$. Check your times tables.

You can also test this by getting out a lot of matchsticks or something else you have a lot of and making twelve groups of 5 or five groups of 12 and counting them all up. How many do you have? 60.

The distributive law also means that if I said to you, "Hey! What is 5×2 plus 5×3?" You could just say, "Please *try* to challenge me a little bit. That's obviously a fancy way to say 5 times 5, which I know is 25."

$$(5 \times 3) + (5 \times 2) = 15 + 10 = 25$$

Another law of multiplication is the **commutative law**. You already know it—really. Remember back a few lines when we said you could count out five groups of 12 or twelve groups of 5? Remember when you memorized the multiplication tables and the horizontal line for 2 was the same as the vertical line for 2?

	2	**3**	**4**	**5**	**6**	**7**	**8**	**9**	**10**	**11**	**12**
2	4	6	8	10	12	14	16	18	20	22	24
3	6										
4	8										
5	10										
6	12										
7	14										
8	16										
9	18										
10	20										
11	22										
12	24										

Officially, the commutative law of multiplication is this:

$$a \times b = b \times a$$

What it means is $2 \times 5 = 5 \times 2$. And what do these things equal? 10.

So you now know practical math, as well as math theory. Impressive.

A Nifty Time Saver

As a bonus, here's another multiplication pattern that will save you time.

What is 5×10? 50. Now what did you do to the 5? You just added a zero to it. And what is 2×10? 20. Notice anything? Whenever you multiply a number by 10, all you have to do is add a 0 to that number.

Back to the presents.

What if, for whatever strange reason, you had to multiply 54 times 23? Let's say you still had 54 relatives on your holiday gift list and you had major amounts of money. But, you were busy and just didn't have time to be personal and specific and you decided to get each of them the same thing, a $23 sheet and towel set. (We'll just imagine for a moment that you have some strange obsession with linens.) So you need to figure out how much this is going to cost you. Fine. You just pull out your sheet of paper and set up your problem.

$$\begin{array}{r} 54 \\ \times\ 23 \\ \hline \end{array}$$

Use the distributive property.
54 times 3 is 50 times 3 plus 4 times 3, or 162.
The other part of your problem now is 54 times 20.
54 times 20 is 50 times 20 or 1,000.
Plus 4 times 20 or 80.
1,000 plus 80 equals 1,080.

$$\begin{array}{r} 162 \\ +\ 1{,}000 \\ +\ \quad 80 \\ \hline 1{,}242 \end{array}$$

The other way to look at this is to look at the columns.

$$\begin{array}{r} 54 \\ \times\ 23 \\ \hline \end{array}$$

Multiply the 4 times the 3 and get 12. The same as with addition, you need to carry the 1 from the 12 over to the next digit. Multiply 5 times 3 and get 15, add the 1 you carried from 12, and it is 16.

$$\begin{array}{r} \overset{1}{5}4 \\ \times\ 23 \\ \hline 162 \end{array}$$

Now you multiply by the tens digit. When multiplying by the tens digit, the product will be stacked below them starting in the tens place, not the units place. Multiply 4 times 2 and get 8. Multiply 5 times 2 and get 10.

$$\begin{array}{r} 54 \\ \times\ 23 \\ \hline 162 \\ +108 \\ \hline \end{array}$$

Miraculous! The same darn thing.

$$\begin{array}{r} 54 \\ \times\ 23 \\ \hline 162 \\ +108 \\ \hline 1{,}242 \end{array}$$

You're also using your addition skills, which are really sturdy by now, so you can count this as addition practice, too.

APPROXIMATE THIS:

3 times 51?
How about 22 times 51?

Now try these multiplication problems to hone your skills. Try using the distributive law, breaking down the harder numbers.

Quiz #8

1. 3 × 17
2. 45 × 12
3. 2 × 41
4. 132 × 195
5. 43 × 1
6. 77 × 11
7. 57 × 0
8. 398 × 43

FACTORS

There are some other things you might want to think about that have to do with multiplication. You know that $2 \times 3 = 6$. But did you also know that both 3 and 2 are **factors** of 6? A factor of a number is any number that can be multiplied by another number to get that first number. Since you can multiply 2 by some other number and get 6 as a product, 2 is a factor of 6. The factors of 6 are: 1, 2, 3 and 6.

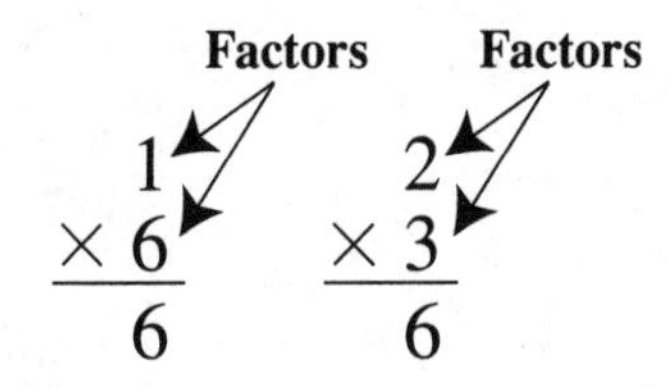

How about the factors of 18? 1, 2, 3, 6, 9 and 18.

There are some numbers so huge it would take forever to list all their factors. And there are other numbers that have only two factors in the world, 1 and the numbers themselves. These numbers are called **prime** numbers. For instance, list the factors of 2: 1 and 2. And that's it. So 2 is a prime number. Check out some other numbers. Is 3 prime? Is 4?

Yes, 3 is prime, $1 \times 3 = 3$. There are no other factors of 3. No, 4 is not prime: $1 \times 4 = 4$ and $2 \times 2 = 4$ so there are three factors of 4: 1, 2, and 4.

Prime numbers are positive.

There's one crazy thing about prime numbers. 1 is not a prime number. Why not? Because some mathematician said so.

APPROXIMATE THIS:

How many numbers between 1 and 100 do you think are prime?

MULTIPLYING NEGATIVE NUMBERS

Remember how you used the idea of owing money as a way of understanding negative numbers?

Look at it this way: you owed money, say $3, to 5 people, it would be -3×5 and you would be $15 in debt. In other words, you'd have what we have been calling -15.

Look at it the same way you looked at regular multiplication. If you have -5×3 you have -5 added three times or $-5 + -5 + -5$. Which you know from your excellent adding negative number skills is -15.

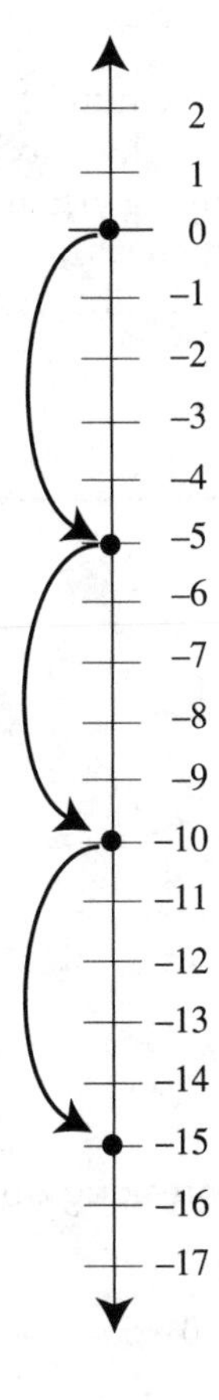

$$-5 \times 3 = -15$$

When you are multiplying a negative number by a positive number, you can treat it almost as though you are multiplying positive numbers. The only difference is that you will stick the negative sign onto the product when you are done.

Multiplying negative numbers by negative numbers is slightly different.

If you DIDN'T have a debt of $3, and you DIDN'T have this debt to 5 people, you would end up being $15 richer than you thought. Subtracting a negative number makes something bigger.

$$-3 \times -5 = +15$$

This is because a negative number times a negative number gives a positive number as the product.

When you are multiplying negative numbers by negative numbers, start by treating them as ordinary numbers and DON'T stick on the negative sign at the end.

Try a few problems:

Quiz #9

1. $-5 \times 7 =$
2. $9 \times -6 =$
3. $-6 \times -5 =$
4. $-5 \times 0 =$
5. $-17 \times -1 =$
6. $-4 \times 72 =$
7. $7 \times 31 =$
8. $-45 \times -67 =$

DIVISION

Division is the process of breaking down numbers into smaller groups.

It can look like this $10 \div 5$, this $5\overline{)10}$, and this $\frac{10}{5}$.

If you see the $\div$ sign between two numbers, the number that comes *before* the $\div$ is being divided by the number that comes *after* the $\div$.

If you see the $\overline{)}$ sign, the number *inside* the box thing is being divided by the number *outside* the box thing.

If you see the $\frac{}{}$ sign, the number on *top* is being divided by the number on the *bottom*.

If you had 40 chocolate bars and you wanted to separate them into 2 equal groups, one for before dinner and one for after dinner, how many would be in each group? (We're just going to assume here that you are really hungry for something sweet.) There are 20 in each group. What you just did was divide 40 by 2.

How about if you wanted to divide 2 into 2 groups? Or 5 into 5 groups? There would be 1 in each group, right? That's why any number, no matter how gigantic or tiny, divided by itself, is always equal to 1.

How about if you wanted to divide 1,765,432 radios among 1 person. You're thinking, "Among 1 person? Is she crazy? You can't divide among one person, the one person would get all of them." Exactly right, you would have to give the one person all 1,765,432 radios. That is why any number divided by 1 is equal to itself.

Division is useful. Think about it. You might need to know how much each person will get if you divide your $100 among your 5 family members. What you are asking in that case is how many times 5 will go into 100. Like all the math you've been doing, you are going to do it piece by piece.

$$5\overline{)100}$$

First you ask yourself, "How many 5s are there in the number 1?" Well, there aren't any, 1 is too small, so you move to include the next digit place. Now ask yourself, "How many 5s are there in the number 10?" There are two 5s that make up the number 10, so on top of your division sign, in the tens place since that is the digits place you are working with, you write 2.

$$\begin{array}{r} 2 \\ 5\overline{)100} \end{array}$$

Since there is still a 0 left, ask yourself, "How many 5s are there in 0?" And you figure there are none, so you write 0 above the units place.

$$\begin{array}{r} 20 \\ 5\overline{)100} \end{array}$$

Twenty dollars is the result. That's how much money you could spend on each family member, if you had $100, and 5 family members, and you felt like buying each one a present.

Notice that 5 times 20 is 100 and 20 goes into 100, 5 times. This is because division is the inverse calculation of multiplication, just like subtraction is the inverse calculation of addition.

How about if you had to divide into a more complex number? For instance, what if you had 578 hot dogs, and you had to divide them among 34 screaming kids in a day camp?

$$34\overline{)578}$$

You approach this problem the same way as the simpler-looking problem, but you need to figure this time how many 34s will fit in a 5? Too small, so how many 34s will fit in a 57?

Here is where approximating can make your life a lot easier. If you had two (use 30 because it is easier to approximate with than 34) you would have 60, so it must be less than 2 and you are going to write down 1 in the tens place.

$$\begin{array}{r} 1 \\ 34\overline{)578} \end{array}$$

Below you are going to use your excellent subtraction skills. Write down 34 below the 57. This is because what you are really doing is multiplying 34 by the 1 on top, and sticking it under the 57 to find the difference.

$$\begin{array}{r} 1 \\ 34\overline{)578} \\ \underline{34} \end{array}$$

Now you subtract the 34 from the 57. You get 23.

$$\begin{array}{r} 1 \\ 34\overline{)578} \\ \underline{34} \\ 23 \end{array}$$

And you drop down the next digit of that 578, which is 8.

$$\begin{array}{r} 1 \\ 34\overline{)578} \\ \underline{34} \\ 238 \end{array}$$

Take a guess, how many 34s will fit into 238? There are about three 30s in every hundred, and you have a little more than two 100s here, so you can try a little more than 6, for instance 7. Multiply 7 by 34 and put your result under the 238.

$$\begin{array}{r} 17 \\ 34\overline{)578} \\ \underline{34} \\ \underline{238} \\ 238 \end{array}$$

There is nothing left over when you divide this time, which means it has divided in evenly. There will be no leftover hot dogs. And you don't have any digits left in 578, so you're done. Each screaming kid will get 17 hot dogs. By the way, the number you are dividing into, in this case 578, is called the **dividend**. The number you are dividing by, in this case 34, is called the **divisor**. And the result, the 17 hot dogs each kid is going to get, is called the **quotient**.

Question: Why is it that when you multiply the 17 times the 34 you get how many hot dogs you started with, 578?

Answer: Because just as you saw with the $20 fitting 5 times into 100, division is the inverse of multiplication.

Whenever you have a number that divides evenly into another number, the way you divided 34 into 578 and got the whole number 17 as your quotient, you know that 578 is **divisible** by both 17 and 34.

That means that 17 and 34 are factors of 578. It's all connected. The most convenient thing about this is that now you have a tidy way to check your answers in division.

Multiply the quotient by the divisor and see if you get the dividend; if this works, you are right on target. This also gives you an opportunity to get more practice in perfecting your multiplication skills. Whoopee, right?

APPROXIMATE THIS:

A mysterious man in dark glasses gives you 500 pieces of pizza to distribute among the 75 people who live on your street. How many pieces does each person get?

APPROXIMATE THIS:

If you have 15 errands to do, and you want to take an equal amount of time to do each errand (you're feeling meticulous), and you have 400 minutes, how many minutes do you have to do each errand?

Now you can try a simple division problem, and work on it as it gets more complex.

There are 4 friends over at your house, and you have a box with 36 shirts in it. If you gave everybody an equal number of shirts, how many would each friend get?

Think about it for a second. What are you dividing up? The shirts, right? So the 36 shirts will be your dividend.

And what are you dividing by? The number of groups, which in this case equals the number of friends. How many groups of shirts will there be? There will be 4.

You have $4\overline{)36}$.

How many groups of 4 are there in 36? 9.

Take a minute to remember that division is the inverse of multiplication. Then remember the multiplication tables for 4 times 9. Since your quotient times your divisor will always give you your dividend, knowing your multiplication tables backward and forward will make your life much easier once you are done memorizing.

Okay, you give each friend 9 shirts, no big deal. But say for a minute that you notice that 5 of your favorite shirts were stashed in the pile by someone clearly not looking out for your best interests. Of course you want to keep them, but you still want to divide up the shirts that are left among your friends. What do you do?

As always, one step at a time.

First thing take your favorite shirts out of the pile and put them back in your closet. Then divide what is left among your friends.

36 shirts − 5 favorite shirts = 31 shirts left

Then, divide the number you get by the 4 friends.

$$4\overline{)31}$$

By approximating, you know that each person won't get 9 shirts, right? Because you are starting with fewer shirts than the last time you divided, each person is going to get fewer shirts. Since 9 is too many, try 8.

$$\begin{array}{r} 8 \\ 4\overline{)31} \\ 32 \end{array}$$

But 32 is bigger than 31. So you can't give them 8 each; there aren't enough shirts, there are only 31. Try 7.

$$\begin{array}{r} 7 \\ 4\overline{)31} \\ 28 \end{array}$$

Well, there are enough shirts to give each person 7, but that means there will be some left over. You give away only 28 if you give each of your 4 friends 7 shirts.

$$4 \times 7 = 28$$

If you give away 28 shirts, how many will be left over out of those 31?

$$\begin{array}{r} 31 \\ \underline{-28} \\ 3 \end{array}$$

You have 3 extra shirts. In math you are going to call this the **remainder**. The remainder is what is left over if something doesn't divide evenly. That means you are trying to divide a number by another number that isn't one of its factors (remember factors?). The answer to this shirt question is 7 remainder 3. Generally it is written as 7 r3. If you want to read that out loud say "seven remainder three."

What would you do in real life if you had 3 extra shirts? Probably give them to one of your closest friends, but then you might hurt someone's feelings because how can you really decide who is closest to you in front of everybody? Or you could put them back in your closet and try to forget about them, but then your mom or your dad or your friend or whoever it was that was giving you a hard time about having too many shirts will eventually find them. But in math all you have to know is that there will be some extra. So really in some ways math is easier than the rest of your life.

Something to Think About

The remainder in a division problem is always going to be smaller than the divisor. Why is that, do you think? Look at it this way: when you divided the 31 shirts among the 4 friends, you gave each person 6 shirts, with a remainder of 7.

$$\begin{array}{r} 6 \\ 4\overline{)31} \\ \underline{24} \\ 7 \end{array}$$

If you had 7 as a remainder, you'd have enough to give everyone 1 more shirt. It shows you can make the piles of shirts, or the quotient bigger. That's why the remainder is never bigger than the divisor.

Try a few more division problems. See which ones divide evenly, and which ones have remainders.

Quiz #10

1. $45 \div 3 =$
2. $65 \div 2 =$
3. $43 \div 9 =$
4. $34 \div 2 =$
5. $17 \div 6 =$
6. $87 \div 9 =$
7. $1,789 \div 45 =$
8. $143 \div 4 =$

Here is another interesting thing about remainders. If you divide a number by 2, and there is no remainder—this is just a fancy way of saying that number is divisible by 2—you call that number **even**. And, if you divide a number by 2 and it does have a remainder of 1, you call the number **odd**. An even number is divisible by 2, and an odd number is not divisible by 2.

Think about the number 4, is it divisible by 2? Yes, so it's even.

3? No, so it's odd.

2? Yes. A number is always divisible by itself, because it is always divisible by its factors, and the number itself is always a factor, because you can multiply it by 1.

How about 0? Can you divide 0 by 2? How many groups of 2 are in 0? Well, there are 0 groups of 2 in 0. If you have 0 divided by 2 is 0, there is no remainder. So 0 is divisible by 2; 0 is even. By the way, 0 divided by any number is always 0, so 0 is divisible by all integers EXCEPT ITSELF. But, if you ever try to divide a number by 0 you just can't do it. You can try it on a calculator and watch it come up ERROR ERROR ERROR.

Dividing Negative Numbers

Dividing negative numbers is the same as multiplying negative numbers, only backward. Remember, division is the inverse of multiplication. Here's what you need to know about dividing negative numbers.

If you divide a negative number by a positive number, or a positive number by a negative number, the result will be NEGATIVE.

$$12 \div (-2) = -6$$
$$-30 \div 3 = -10$$

If you divide a negative number by another negative number the result will be POSITIVE.

$$-15 \div -3 = 5$$

Take a look at a problem that combines these ideas.

$$2 \times -6 \div 3$$

First you multiply, since that comes first in the problem. You have a positive multiplied by a negative, so the product is negative.

$$2 \times -6 = -12$$

Now you have division. A negative divided by a positive, so the quotient will be negative.

$$-12 \div 3 = -4$$

Your final answer is -4.

Try some combination multiplication and division problems on your own. Some of them have both negatives and positives.

Do these step-by-step, only two numbers at a time, in the order the numbers appear.

Quiz #11

1. $3 \times 4 \div 6 =$
2. $7 \times 8 \div -7 =$
3. $9 \times (-8) \times 7 \div -5 =$
4. $15 \div -5 \times -3 =$
5. $6 \div 3 \times 7 =$
6. $-34 \div 2 \times 142 \div -3 =$
7. $156 \div 18 =$
8. $72 \div -6 \times -5 =$

Okay, take a look at your answers and make sure you feel okay about these problems, because now starts what you have been waiting for: COMBINING EVERYTHING SO FAR.

Before combining what you've learned so far you'll need to learn one last concept. Then you'll need to learn how to put everything in order.

The concept you need to get cozy with is the idea of the **exponent**. Exponents are sort of shorthand ways for people to write out long boring multiplication problems that include the same number over and over and over. For instance:

$$3 \times 3 \times 3 \times 3 \times 3$$

can be written as 3^5 using exponents. The little floating 5 up there on the right is the exponent, and it shows that you are going to multiply the big 3 on the bottom by itself 5 times. To read 3^3 is 243. And 4^3 is $4 \times 4 \times 4$ which equals 64. When you put an exponent, like 3 in this case, with a number, like 4 in this case, it is called **raising** 4 to to the third power.

There are names of the most commonly used powers, 2 and 3.

$$4 \times 4$$

This is 4^2, four to the second power. It's also called four **squared**.

$$4 \times 4 \times 4$$

This is 4^3, four to the third power. It's also called four **cubed**.

When you square a negative number, it becomes positive. That's because a negative number times a negative number gives a positive product, remember? So the square of any number in the world is a positive number. Even if the number is –999,999,999, its square will be positive. In fact, if you raise any number to an even power it will give a positive product for the same reason.

$$-2^2 = -2 \times -2 = 4$$

When you raise a negative number to an odd power, the product will be negative, because a negative times a positive gives you a negative.

$$-2^3 = -2 \times -2 \times -2 = -8$$

When you raise 1 to any power, the result is always 1. It makes sense.

$$1 \times 1 \times 1 \times 1 \times 1 \times 1 \times 1 \times \ldots = 1$$

Any number raised to the first power, like 5^1, is equal to that number, in this case, 5.

Anything raised to the 0 power is equal to 1. It's a rule.

And for numbers that people use a lot, like 10, the results of 10 being raised to the 2nd power (100), and the 3rd power (1,000), and all other exponents, are called the **powers of 10.**

To feel cool, calm, and in control, try a few of these:

Quiz #12

1. 4^3
2. 3^6
3. -2^4
4. 5^0
5. -4^2
6. 1^7
7. -3^3
8. 7^1

So. You are happy with exponents. A few of you might now be saying, "What about square roots?" We are going to put the rest of exponents, which includes square roots, off for a while. Now we need to talk about the Order of Operations.

THE ORDER OF OPERATIONS

If you see a complicated-looking math problem, how do you stay calm? You do the calculations one at a time. And you don't just go from left to right, the way you read the problem. Instead you follow this order.

PARENTHESES EXPONENTS MULTIPLICATION DIVISION ADDITION SUBTRACTION

A lot of people remember this order using mnemonics like:

PEMDAS
or
Please Excuse My Dear Aunt Sally

Whichever way you want to think about it is okay as long as you remember the order of operations.

Parentheses; do what is inside of these first. Parentheses look like this: ( ). They're a little ridiculous by themselves, but when they surround things, watch out.

Solving the part inside the parentheses first,

$6+7+8\div 2(3+2)\times 3^2-6+3$, since (3 + 2) = 5, becomes

$6+7+8\div 2(5)\times 3^2-6+3$

In any arithmetic problem, do the part inside the parentheses first!

Exponents, if there are any, get done next.

$6+7+8\div 2(5)\times 3^2-6+3$, since $3^2 = 9$, becomes

$6+7+8\div 2(5)\times 9-6+3$

Once you have gotten through the parentheses and exponents, multiplication and division get done next. Their order is interchangeable, so go left to right.

$6+7+4(5)\times 9-6+3$, since $8\div 2$ is 4, 4×5 is then 20, and 20×9 is 180, becomes $6+7+180-6+3$

By the way, parentheses next to a number, like 4(5), mean multiply by, just like 4×5.

Addition and subtraction are last. Do these left to right in the order you find them, just like multiplication and division.

$$6 + 7 + 180 - 6 + 3 = 190 \text{ because}$$
$$6 + 7 + 8 \div 2(3 + 2) \times 3^2 - 6 + 3 = 190$$

Beware! Calculators don't have a specific memory for the order of operations so they may calculate incorrectly if you just punch in the numbers as you see them in a problem.

You are ready to deal with just about anything now.

APPROXIMATE THIS:

How many paperback books will fit in a standard-sized suitcase?

Oh, and by the way, you, who were scared of all this math, have now successfully mastered the underpinnings of mathematics. Congratulations.

The following glossary is to provide you with the exciting opportunity to review the definitions of terms that may be new or unfamiliar to you from this chapter.

GLOSSARY

integers: All real numbers other than decimals or fractions.

positive number: Any number greater than 0.

negative number: Any number less than 0.

whole number: all positive integers and 0.

∞: Infinity.

absolute value: The number of spaces a number is from 0 on the number line. The absolute value of a number is always positive.

approximation: The process of guessing the rough size or amount of something rather than calculating it exactly.

digits: The integers 0 through 9.

sum: The result of addition.

product: The result of multiplication.

distributive law: Multiplying $a \times (b+c)$ is equal to multiplying $a \times b$ and adding it to $a \times c$.

commutative law: Multiplying $a \times b$ is equal to multiplying $b \times a$.

factors: The factors of some particular number are the numbers that multiplied together, form the product in question.

prime: A prime number has only two factors, itself and 1. Don't forget, 1 is not a prime number.

dividend: The number being divided into in a division problem.

divisor: The number that you're dividing by in a division problem.

quotient: The result of division.

divisible: The quality of being able to be divided with no remainder.

remainder: The number left over when a number is not evenly divisible.

even: Divisible by 2.

odd: Not divisible by 2.

exponent: A small raised number that indicates how many times to multiply the big number times itself.

power: The numerical amount of an exponent.

squared: A number raised to the second power is being "squared."

cubed: A number raised to the third power is being "cubed."

powers of ten: The results of 10 being raised to any exponential power.

Try these problems to test your math skills:

Quiz #13

1. $8^2 - (7 \times 6 \times 5) - (3 - 1) \div 2 =$

2. Your neighbor has 15 meowing cats, 7 rock bands with 4 musicians each, and 6 dogs living with him. After you complain for a year, he gets 4 of the cats and 3 of the musicians to move out. If, because you

are so nice, you decide to distribute equally among this insane household 126 long-stemmed roses on Valentine's Day, how many roses would each person or animal receive? (Don't worry, you don't have to count the neighbor himself.)

3. Oh my God! You have to get each of your 13 brothers and sisters a cowboy hat to wear to the embarrassing family barbecue. If each hat costs $8, and you have absolutely no money at all right now, how many hours are you going to have to work at your depressing hellish job where you get paid $4 an hour in order to pay for all those hats? (Don't worry about taxes, please. Your life is hard enough right now.)
4. $9-8-7-6-5-4-3-2-1=$
5. You won a raffle! You received $20,000, and now you have to split it up among your family, your friends, and yourself. You decide you want to keep $12,344. Now you need to divide the rest equally among your mother, your father, your brother, and your friends Joe, Franca, and Serena. Remember, dividing equally means each person gets the same amount. How much will each of them get?
6. $-3\times-4\times-5\times-6=$
7. $-34{,}065\div-45=$
8. Every time you cross Smith Street, the woman on the corner gives you $5 (don't ask why). Every time you cross Seventh Avenue, a guy grabs $2 away from you. If you cross Smith Street 8 times, and you cross Seventh Avenue 10 times, and you started out with no money, how much money, if any, do you end up with?
9. You have 15 errands to run, and each of the errands takes 20 minutes. If you need to be done with your errands by 4 o'clock in the afternoon, what is the latest you can start the errands?
10. $3+(-3\times-4)-4^2\times5(4-3)\div4=$

ANSWER KEY

Quiz #1

1. 468
2. 1,101
3. 683
4. 200
5. 502
6. 111
7. 386
8. 13,999

Did you try to add 7,000 and 7,000 and then get rid of the 1?

Quiz #2

1. 66
2. 30
3. 56
4. 1
5. 808
6. 98
7. 19
8. 1,995

Quiz #3

1. –1
2. –14
3. –1
4. 11
5. –15
6. 6
7. –41
8. –28

Quiz #4

1. –1
2. 1
3. 9
4. –4
5. 7
6. 14
7. 46
8. 32

Quiz #5

1. 11
2. 5
3. –35
4. –6
5. –5
6. –18
7. 47
8. 10

Quiz #6

1. 47
2. 8
3. –94
4. 10
5. –29
6. 24
7. 3
8. 51

Quiz #7

1. 6 hamburgers
   How It Works:
   The hamburgers are your numbers, so $3-2$ is when John has 3 hamburgers and gives 2 to Cheryl.
   $3-2=1$
   $1+3+5=9$
   $9-3=6$
   The 3 and the 5 are from Bob and Gregory, then the subtracted 3 are those that John gives away to his brother Bill. So John ends up with 6 hamburgers.

2. 18 hats
   How It Works:
   The hats here are being counted.
   $15\ (\text{start}) - 5\ (\text{given away}) = 10$
   $10 + 12\ (\text{new shipment}) = 22$
   $22 - 7\ (\text{thrown away}) = 15$
   $15 + 3\ (\text{sidewalk sale}) = 18$
   She has 18 hats at the end of the day.

3. \$19
   How It Works:
   Starts with –25 (he owes his boss 25)
   $50\ (\text{from Julie}) + -25 = 25$
   (he paid his boss back)
   $25 - 6\ (\text{imported beer}) = 19$
   He has \$19 after paying back the boss and buying the beer.

4. No more time left.
   How It Works:
   Start with 20 minutes.
   $20 - 7\ (\text{friend talk}) = 13$
   $13 - 2\ (\text{candy bar}) = 11$
   $11 - 11\ (\text{space out}) = 0$
   Time to be in class is now! You have no more time left.

5. 3 dishes left.
   How It Works:
   You start out with all the dirty dishes, 12.
   $12 - 2\ (\text{Mike}) = 10$
   $10 - 3\ (\text{thrown out}) = 7$
   $7 - 4\ (\text{you washed them}) = 3$
   You still have 3 dishes left. What do you think, would it be easier to add up the number of dishes washed and subtract that from the number of dishes you start with? Try it and see.

6. 3,094 jellybeans in the safe.
   How It Works:
   She starts with 3,241.
   $3{,}241 - 3\ (\text{tonight}) = 3{,}238$
   $3{,}238 - 20\ (\text{tomorrow}) = 3{,}218$
   $3{,}218 - 53\ (\text{next day}) = 3{,}165$
   3,165 – 71 (the day after that) = 3, 094
   She has 3,094 jellybeans in the safe.

7. Jennifer owes Alice $73.
How It Works:
Jennifer starts owing Alice $500, or we can look at it as though she has $-\$500$.

$-500 + 275 \text{ (paycheck)} = -225$

$-225 + 35 \text{ (parents)} = -190$

$-190 + 117 \text{ (raffle prize)} = -73$

So she still owes Alice $73, or look at it as though Jennifer has $-\$73$.

8. He's 5 minutes late.
How It Works:
George started with 180 minutes.

$180 - 25 \text{ (gym)} = 155$

$155 - 30 \text{ (girlfriend)} = 125$

$125 - 80 \text{ (pool hall)} = 45$

$45 - 50 \text{ (lunch)} = -5$

He's late! He's 5 minutes late because he's dealing with negative time here. Would it be easier here to add up the time he spent and then subtract it from the whole? How many different ways are there to do these? Give yourself some time, try it out.

Quiz #8

1. 51
2. 540
3. 82
4. 25,740
5. 43
6. 847
7. 0
8. 17,114

Quiz #9

1. –35
2. –54
3. 30
4. 0
5. 17
6. –288
7. 217
8. 3,015

Quiz #10

1. 15
2. 32 r1
3. 4 r7
4. 17
5. 2 r5
6. 9 r6
7. 39 r34
8. 35 r3

Quiz #11

1. 2
2. –8
3. 100 r4
4. 9
5. 14
6. 804 r2
7. 8 r12
8. 60

Quiz #12

1. 64
2. 729
3. 16
4. 1
5. 16
6. 1
7. –27
8. 7

Quiz #13

1. –147
2. Each household member gets 3 roses.

   How It Works:

   15 (cats) + (7 × 4) (musicians) + 6(dogs) =49

   49 – (4(cats) + 3(musicians) = 42

   126(roses) ÷ 42 = 3

   Three roses for each household member.
3. You need to work 26 hours.

   How It Works:

   13 (people) ×8($ per hat) = 104

   104 (hat $) ÷ 4 (your earnings) = 26

   You need to work 26 hours to pay for all the hats. How do you know you need to divide here? Think of it as all that money being split up into hour-by-hour chunks of 4 dollars each, and you find out how many of those hours you will need.
4. –27
5. Each person gets $1,276.

   How It Works:

   Get rid of your money: $20,000 - 12,344 = 7,656$

   Now split it up, better known as divide.

   Each person gets $1,276.
6. 360
7. 757
8. You make $20.

   How It Works:

   (8 ×5)Smith Street Seventh + (10 × –2) Seventh Avenue = 20

   You make $20.

9. You need to start by 11 o'clock that morning.
   How It Works:
   $15 \times 20 = 300$ minutes
   $300 \div 60$ (minutes in each hour) $= 5$
   If it will take you 5 hours, you need to start by 11 o'clock. Draw a picture of a clock and see. Do you need to convert it into hours? Not really, just count 60 minutes for every time you go around the clock face. And remember to go backward on the clock, because you are seeing when to begin.
10. –5

CHAPTER 2

FRACTIONS

Oh my God it's fractions! Does that make you want to run for your life? Relax--hold on for a while and see what is ahead. Things won't seem as frightening as they do now. And who knows, they may even start to make sense.

Say I have this apple. Only 1 apple, and all the stores are closed so I can't buy any more apples. You and I are sitting next to each other in a park, and I see you eyeing my apple. You're hungry, and I mean really hungry. Finally, because you won't let up, I say, "Fine, you can have some of the apple. Okay? Are you happy now?" And I hand over the apple to you and you take a big bite. What you just did was take a **fractional** part of the apple. A part out of the whole. We divided the apple, even though we may not have divided it evenly.

APPROXIMATE THIS:

Three bites is what fractional part of an apple?

Fractions are most easily recognized when they use the fraction bar. The fraction bar, that line between the 2 numbers, shows that division is taking place.

$$\frac{6}{3}$$

In some cases the numbers will divide evenly, like $\frac{6}{3}$. In other cases, however, these numbers will not divide evenly, like $\frac{5}{2}$, or $\frac{2}{3}$. They leave remainders.

All these types of numbers are called **fractions** whether they express division or part over whole. Fractions are actually pretty great.

A fraction is made up of two parts: the top, also called the **numerator**, and the bottom, also called the **denominator**. In the fraction $\frac{5}{2}$, 5 is the numerator, and 2 is the denominator. The numerator represents the part, and the denominator represents the whole.

For positive numbers, when the numerator (top) of the fraction is a number larger than the denominator (bottom), like $\frac{5}{2}$, the value of the fraction is greater than 1. You already know that because you know that $\frac{5}{2} = 5 \div 2 = 2\text{ r}1$. People in math call this an **improper fraction**. The part is 5 and the whole is 2. Think of it as cutting up apples in 2 pieces, and having 5 of these pieces. Five halves is more than 1 apple.

For positive numbers, when the numerator of the fraction is smaller than the denominator, like $\frac{2}{3}$, the value of the fraction is less than 1. This is because 3 can't fit into 2 even one time. This is called a **proper fraction**. It's like cutting up an apple into 3 pieces, and only having 2 of these parts.

And remember dividing a number by itself? What if you cut an apple into 3 pieces and you had all 3 of them? You would have the whole apple, wouldn't you.

When the numerator and the denominator have the same value, like $\frac{3}{3}$, the fraction is exactly equal to 1.

What about when your fraction is a number over 1? Any number over 1 is equal to that number. For instance, $\frac{3}{1}$ equals 3. Think about if you had an apple divided into 1 piece; in other words, a whole apple. What if you had 3 of them? Exactly as it seems; you would have 3 apples.

Remember, you can never divide 0 into any number, so the denominator of a fraction can never be 0.

What about when you divide 0 by something else? Well a fraction with the numerator 0 is always equal to 0, no matter what number it has for a denominator. If you had 0 apples cut up into 3 pieces, how many sets of pieces? 0.

Fractions Smaller Than 1

Lots of math deals with fractions that are smaller than 1. We could think back to the apple and figure that you took a small bite, and maybe the apple consisted of 5 more bites exactly that size, or 6 bites total. So the bite you took was $\frac{1}{6}$, or 1 bite out of a possible 6. The 1 bite represents the part of the apple, and the 6 bites represent the whole. It's back to what we said before—fractions represent part over whole. You say, "one sixth." This number is less than 1, because it is less than the whole of the apple.

Here's another example: Let's say you have a box of 100 pencils. These 100 pencils represent the whole—the whole box of pencils. If you gave someone 20 of the pencils, you would be giving them $\frac{20}{100}$ of the pencils. The numerator, 20, represents the part, and the denominator, 100, represents the whole. The fraction $\frac{20}{100}$ is smaller than 1, but it still represents 20 out of 100 pencils. In most math, $\frac{20}{100}$ is then reduced to a simpler fraction.

Reducing Fractions

Reducing a fraction means dividing *both* the numerator and the denominator by the same number to get a simpler fraction. Use the same fraction we just saw, $\frac{20}{100}$. Look at both the numerator and the denominator. What is a factor of both these numbers? They are both even, so 2 will divide evenly into both of them.

Divide the 20 by 2, and the 100 by 2, and get $\frac{10}{50}$

$$\frac{20 \div 2}{100 \div 2} = \frac{10}{50}$$

Since both 10 and 50 still have a common factor, you can reduce again.

$$\frac{10 \div 10}{50 \div 10} = \frac{1}{5}$$

This fraction, $\frac{1}{5}$, is the simplest form of the same fraction. You know this because 1 has no other factors, so you couldn't divide it again. By giving someone 20 pencils, you are giving them $\frac{1}{5}$ of the box of pencils. Fractions can look entirely different. For instance, $\frac{20}{100}$ and $\frac{1}{5}$ and $\frac{4}{20}$ and $\frac{3}{15}$ still represent the same number. All the fractions in the last sentence are equal.

What fraction of the 100 pencils would you be giving away if you gave away 50 pencils?

What fraction of the 100 pencils would you be giving away if you gave away 45 pencils?

What fraction of the 100 pencils would you be giving away if you gave away 27 pencils?

Try to reduce these fractions as well:

Quiz #1

1. $\frac{4}{6}$
2. $\frac{12}{144}$
3. $\frac{5}{9}$
4. $\frac{3}{9}$
5. $\frac{15}{30}$
6. $\frac{39}{90}$
7. $\frac{12}{60}$
8. $\frac{9}{9}$

Look at that last fraction. What if it had been $\frac{5}{5}$? $\frac{7}{7}$? $\frac{2}{2}$? $\frac{-4}{-4}$? Remember, any number divided by itself is equal to 1.

WORKING WITH FRACTIONS

Fractions make some sense now, right? Because if you didn't have fractions to describe things, if someone asked you for some pie you would have to give them the whole pie because there would be no good way to say, "Do you want $\frac{1}{2}$ of the pie? Or $\frac{1}{8}$?"

Which of the two fractions, $\frac{1}{8}$ or $\frac{1}{2}$ would give you more pie? Which of these is the bigger fraction?

Think about it as getting a pie at the bakery and cutting it up.

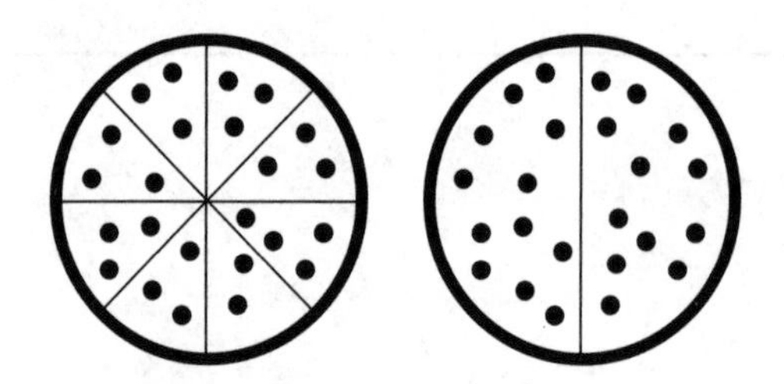

If you cut the pie into 8 slices, you get smaller slices than if you cut the pie into 2 gigantic slices. So $\frac{1}{8}$ is smaller than $\frac{1}{2}$. Here's an even easier way to compare fractions so you don't have to draw pictures of pies at awkward moments. Call it the bow tie.

BOW TIE

$$\frac{3}{5} \quad \frac{5}{6}$$

Try multiplying the denomimator (remember the number on the bottom is the denominator) of the fraction on the right by the numerator (the number on top) on the left and put the product above the fraction on the left.

$$\overset{18}{\frac{3}{5}} \nwarrow \frac{5}{6}$$

Now, multiply the denominator of the fraction on the left by the numerator of the fraction on the right and put that product over the fraction on the right.

$$\overset{18}{\frac{3}{5}} \times \overset{25}{\frac{5}{6}}$$

Now which is bigger, 18 or 25? 25, so the fraction under the 25 is the bigger fraction. If someone offers you $\frac{3}{5}$ or $\frac{5}{6}$ of a million dollars, take the $\frac{5}{6}$.

You might be wondering how this works. Read on, and the more you know about fractions the more sense the bow tie will make.

ADDING FRACTIONS

Think for a minute about what it is you are adding when you add fractions. Let's say you have the same box of pencils and you give Julie 3 pencils, or $\frac{3}{100}$ of the box, and you give Sam 1 pencil, or $\frac{1}{100}$ of the box. How many pencils have you given away total? 4. Out of how many pencils? 100. So, you've given away $\frac{4}{100}$ of the box of pencils.

$$\frac{3}{100}+\frac{1}{100}=\frac{4}{100}$$

You add the numerators, but not the denominators. The denominator, or the whole, does not change. You are still talking about 100 pencils. Only the number given away has been altered.

But what happens when the denominators of the fractions are different? If, for example, you want to combine $\frac{1}{2}$ an apple with $\frac{2}{5}$ another apple, you need to add the fraction $\frac{1}{2}$ to the fraction $\frac{2}{5}$.

$$\frac{1}{2}+\frac{2}{5}$$

How do you add $\frac{2}{5}$ and $\frac{1}{2}$? By changing the denominators so they are the same. This is called finding a **common denominator**.

A common denominator is a number that is divisible by both denominators of the fractions you are adding. In this case, since the denominators of the two fractions you are working with are 2 and 5, you can choose the number 10: 10 is divisible by both 2 and 5. The best common denominators are the numbers that are easiest to work with. For instance, 50 is a common denominator because it is also divisible by 5 and 2, but 10 is a smaller number, so it is easier to use.

To change the fraction $\frac{1}{2}$ into a fraction with a denominator of 10, multiply the fraction by $\frac{5}{5}$. Remember, any number over itself is equal to 1, and you can multiply any number by 1 without changing the number. Since $\frac{5}{5}$ is equal to 1, you don't change the value of the fraction. It's like reducing in the opposite direction.

$$\frac{1}{2} \times \frac{5}{5} = \frac{5}{10}$$

To make $\frac{2}{5}$ into a fraction with a denominator of 10, multiply by $\frac{2}{2}$.

$$\frac{2}{5} \times \frac{2}{2} = \frac{4}{10}$$

Now you can add $\frac{4}{10}$ and $\frac{5}{10}$ the same way you added the fractions where the denominators started out the same.

$$\frac{4}{10} + \frac{5}{10} = \frac{9}{10}$$

You've got $\frac{9}{10}$ of an apple when you combine them.

Finding a common denominator works for fractions larger than 1 in exactly the same way. Try adding $\frac{5}{2}$ and $\frac{4}{3}$.

$$\frac{5}{2}+\frac{4}{3}$$

The easiest to use common denominator of 2 and 3 is 6.

$$\frac{5}{2}\times\frac{3}{3}=\frac{15}{6} \text{ and } \frac{4}{3}\times\frac{2}{2}=\frac{8}{6}$$

You have $\frac{15}{6}+\frac{8}{6}=\frac{23}{6}$.

One way to find a common denominator is to multiply the denominators of a fraction addition problem together. This way, you can be sure that your common denominator is divisible by both.

Try these:

Quiz #2

1. $\frac{1}{7}+\frac{1}{2}=$
2. $\frac{1}{8}+\frac{3}{4}=$
3. $\frac{10}{9}+\frac{1}{22}=$
4. $\frac{2}{13}+\frac{1}{10}=$
5. $\frac{3}{5}+\frac{6}{7}=$
6. $\frac{2}{3}+\frac{1}{13}=$
7. $\frac{14}{5}+\frac{11}{2}=$
8. $\frac{5}{6}+\frac{1}{4}=$

Things to Think About

Have you figured out why the bow tie method on page 66 works when you are comparing fractions? The bow tie creates a common denominator. Take a look back at those fractions you were comparing, $\frac{3}{5}$ and $\frac{5}{6}$. Try to add them.

$$\frac{3}{5}+\frac{5}{6}$$

To get a denominator of 30, you would multiply $\frac{3}{5}$ by $\frac{6}{6}$.

$$\frac{3}{5}\times\frac{6}{6}=\frac{18}{30}$$

And you would multiply $\frac{5}{6}$ by $\frac{5}{5}$.

$$\frac{5}{6}\times\frac{5}{5}=\frac{25}{30}$$

So now you have, by the way, $\frac{18}{30}+\frac{25}{30}=\frac{43}{30}$.

It's easy to compare them this way: 18 parts out of a possible 30, or 25 parts out of a possible 30; 25 will give you more so it's bigger. The bow tie just skips writing down the common denominator. If you find out enough, math shortcuts make a lot of sense.

LOWEST COMMON DENOMINATOR

Sometimes in math, people use a special type of common denominator called the **lowest common denominator**. Look at it this way: you have $\frac{3}{4}$ of a box of chocolates and your sister has $\frac{1}{6}$ of a box of chocolates. You're trying to figure out if together you have 1 whole box of chocolates to give to your mother on Mother's Day. So you want to add the two fractions, $\frac{3}{4}$ of a box and $\frac{1}{6}$ of a box.

$$\frac{3}{4}+\frac{1}{6}$$

As you know, you want to find a common denominator to solve this problem. Of course, one way to find the common denominator would be multiply the 2 original denominators together: $4 \times 6 = 24$

$$\frac{18}{24}+\frac{4}{24}=\frac{22}{24}$$

Reduce $\frac{22}{24}$ and you get $\frac{11}{12}$. You and your sister don't quite have 1 whole box. You are a little bit short.

Here's another way to figure out the same problem. Your denominators, 4 and 6, are factors of 24, but they are also factors of a smaller number: 12. How would you know to look for 12? Try fooling around with the larger denominator. First see if 6 could serve as the denominator. It can't, because 4 is not a factor of 6. Try multiplying 6 by 2. You get 12. Well, what about 12? Sure, because 4 and 6 are both factors of 12.

Multiply $\frac{3}{4}$ by $\frac{3}{3}$ to get 12 as your denominator.

$$\frac{3}{4} \times \frac{3}{3} = \frac{9}{12}$$

Multiply $\frac{1}{6}$ by $\frac{2}{2}$ to get 12 as your denominator.

$$\frac{1}{6} \times \frac{2}{2} = \frac{2}{12}$$

Now add.

$$\frac{9}{12} + \frac{2}{12} = \frac{11}{12}$$

Because fractions can look different and mean the same thing, this is the same addition problem as $\frac{18}{24} + \frac{4}{24} = \frac{22}{24} = \frac{11}{12}$. Same answer, smaller denominator to start, and you skip the step of reducing. Which is the better way to do it, lowest common denominator or regular old common denominator? Whichever one you feel more comfortable using.

Mixed Numbers

Some fractions are presented as **mixed numbers**. A mixed number is a number that contains both an integer and a fraction, like $4\frac{2}{3}$. You can look at $4\frac{2}{3}$ as representing 4 cans of peas with an additional $\frac{2}{3}$ of a can of peas.

One of the easier ways to work with a number like this is to treat it as an addition problem.

First turn the integer 4 into a fraction, $\frac{4}{1}$. Then $4\frac{2}{3}$ looks like $\frac{4}{1}+\frac{2}{3}$.

Since $4\frac{2}{3}$ has turned into an addition problem, you need to find a common denominator. Since $\frac{4}{1}$ has a denominator of 1, the denominator of $\frac{2}{3}$, which is 3, will work as a common denominator since it is a multiple of 1. Now, just like in any other fraction addition problem, multiply the fraction $\frac{4}{1}$ by $\frac{3}{3}$.

$$\frac{4}{1}\times\frac{3}{3}=\frac{12}{3}$$

Now you can add.

$$\frac{12}{3}+\frac{2}{3}=\frac{14}{3}$$

So you see, $\frac{14}{3}$ is the same number as $4\frac{2}{3}$. You have converted a mixed number into a regular fraction.

QUICK WAYS TO CONVERT MIXED NUMBERS

A faster way to convert mixed numbers is to multiply the denominator of the fraction, in this case 3, by the integer, in this case 4.

$$4 \nwarrow \frac{2}{3}$$

Then add this product to the numerator of the fraction, in this case, 2.

$$^{12}4 \nwarrow \frac{2}{3}=\frac{14}{3}$$

Put this over the original number.

$$\frac{14}{3}$$

You now have a fraction to work with, instead of a mixed number.

Convert these mixed numbers to regular fractions:

Quiz #3

1. $3\frac{1}{3}$
2. $13\frac{2}{3}$
3. $5\frac{5}{6}$
4. $10\frac{3}{7}$
5. $6\frac{1}{2}$
6. $1\frac{1}{2}$
7. $3\frac{4}{5}$
8. $7\frac{3}{7}$

To convert fractions back into mixed numbers, start by dividing the numerator by the denominator.

For instance, $\frac{7}{5}$ is 7 divided by 5. You get 1 (or $\frac{5}{5}$) with a remainder of 2. The 2 is the remaining part out of 5, or the whole. So you have part over whole, or $\frac{2}{5}$, in addition to the

1. $1\frac{2}{5}$.

Just divide the numerator by the denominator and put the remainder over the denominator.

Convert these fractions into mixed numbers:

Quiz #4

1. $\frac{7}{4}$
2. $\frac{9}{2}$
3. $\frac{13}{4}$
4. $\frac{23}{8}$
5. $\frac{34}{5}$
6. $\frac{70}{3}$
7. $\frac{15}{7}$
8. $\frac{12}{5}$

QUICK WAYS TO ADD MIXED NUMBERS

When numbers are in mixed form, like $2\frac{2}{3}$ and $1\frac{1}{4}$, it is sometimes easier to add them in two separate parts. Think back to the cans of peas. How would you figure out how many you had total? By combining the whole cans.

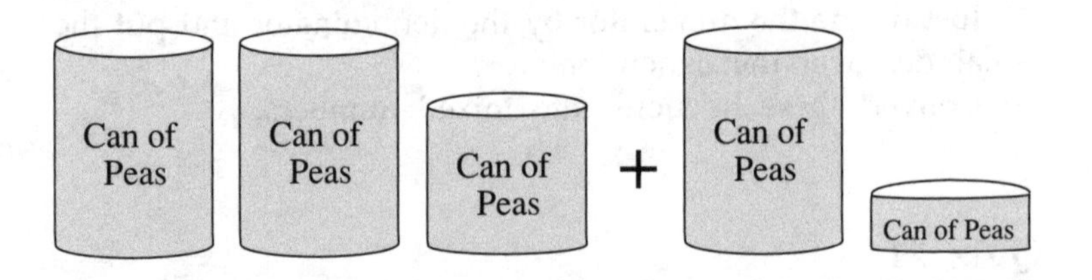

This means to add $2\frac{2}{3}+1\frac{1}{4}$, first add the integers $2+1=3$. Then add the fractions.

$$\frac{2}{3}+\frac{1}{4}=\frac{8}{12}+\frac{3}{12}=\frac{11}{12}$$

Combine the new integer, 3, with the new fraction, $\frac{11}{12}$, and the answer is $3\frac{11}{12}$.

You can then convert this into a fraction.

$3\frac{11}{12}$ becomes

$$12\times3=36$$

$$36+11=47$$

$$\frac{47}{12}$$

That's a lot of fraction information. Make sure you feel comfortable with all these definitions and processes before you keep going. If you have doubts about what fractions mean, just think back to the parts of an apple or a pie cut up into slices or that box of pencils or cans of peas. Feel free to reread. Try adding these fractions before you check out subtraction. Think about which of these problems is easier with any old common denominator, and which was easier with the lowest common denominator.

Quiz #5

1. $\frac{4}{5}+\frac{6}{7}=$

2. $3\frac{3}{4}+5\frac{2}{3}=$

3. $\frac{23}{6}+\frac{3}{12}=$

4. $\frac{2}{5}+3\frac{4}{5}=$

5. $\frac{6}{4}+\frac{9}{4}=$

6. $\frac{91}{4}+\frac{2}{3}=$

7. $\frac{15}{16}+\frac{2}{32}=$

8. $\frac{3}{1}+\frac{4}{1}=$

SUBTRACTION

Subtracting fractions is a lot like adding fractions. You find a common denominator in the same way, and instead of adding the numerators you subtract them. Let's say you have the box of 100 pencils again. You give away (you're getting very generous) 47 pencils, or $\frac{47}{100}$ of the pencils.

$$\frac{100}{100}-\frac{47}{100}=\frac{53}{100}$$

So you now have $\frac{53}{100}$ of the pencils.

Now try subtracting fractions with different denominators. You have $\frac{1}{2}$ of a bag of marbles and you have decided to give $\frac{2}{5}$ of the bag away to your young nephew.

$$\frac{1}{2} - \frac{2}{5}$$

Just like before, when you added fractions, you find a common denominator, in this case 10.

$$\frac{1}{2} \times \frac{5}{5} = \frac{5}{10}$$

$$\frac{2}{5} \times \frac{2}{2} = \frac{4}{10}$$

$$\frac{5}{10} - \frac{4}{10} = \frac{1}{10}$$

The kid gets $\frac{1}{10}$ of a bag of marbles with which to become a marble great.

Subtracting Fractions Greater than 1

You can also subtract fractions that are larger than 1. Say you have $\frac{7}{5}$ of a can of peas this time, and your friend Chuck is taking $\frac{4}{3}$ of a can of peas from you.

$$\frac{7}{5} - \frac{4}{3}$$

The common denominator (by multiplying the two denominators together) is 15.

$$\frac{7}{5} \times \frac{3}{3} = \frac{21}{15}$$

$$\frac{4}{3} \times \frac{5}{5} = \frac{20}{15}$$

$$\frac{21}{15} - \frac{20}{15} = \frac{1}{15}$$

You are left with $\frac{1}{15}$ of a can of peas.

APPROXIMATE THIS:

If you knew $\frac{1}{8}$ of all the people in the world and you invited all of them for dinner, would you be able to fit them on your block even? And how many people is that, anyway?

ANOTHER THING TO THINK ABOUT

Remember when you subtracted a larger number from a smaller one? Like $5-7$? You just said, Okay, the easiest way to figure it is to subtract the 5 from the 7 and stick on the negative sign, and that is going to be -2. Well, the same method works with fractions, too.

$$\frac{1}{3}-\frac{1}{2}$$

Find a common denominator, 6, for instance, and you have:

$$\frac{2}{6}-\frac{3}{6}$$

Now you can work with the numerators. Since 3 is bigger than 2, subtract 2 from 3, and don't forget to stick on the negative sign.

$$\frac{2}{6}-\frac{3}{6}=-\frac{1}{6}$$

How About Adding and Subtracting Negative Fractions?

Owing someone $\frac{1}{2}$ of a dollar is having $-\frac{1}{2}$ of a dollar.

Imagine you owed Fred $\frac{1}{4}$ of a dollar and Martha $\frac{1}{2}$ of a dollar.

$$-\frac{1}{4}+-\frac{1}{2}$$

$$-\frac{1}{4}+-\frac{2}{4}=-\frac{3}{4}$$

All together, you have $-\frac{3}{4}$ of a dollar, or you owe $\frac{3}{4}$ of a dollar.

And if you owe your mother $\frac{3}{4}$ of a dollar, and she says to forget about $\frac{1}{4}$ of the dollar debt—mothers are like that—how much money, or negative money, do you have now?

Remember, subtracting a negative number is just like adding a positive number.

$-\frac{3}{4}-\left(-\frac{1}{4}\right)$ is going to be

$$-\frac{3}{4}+\frac{1}{4} \text{ or } -\frac{2}{4}$$

Can you reduce $-\frac{2}{4}$? Of course. $-\frac{2\div 2}{4\div 2}=-\frac{1}{2}$

Take a look at it on the number line.

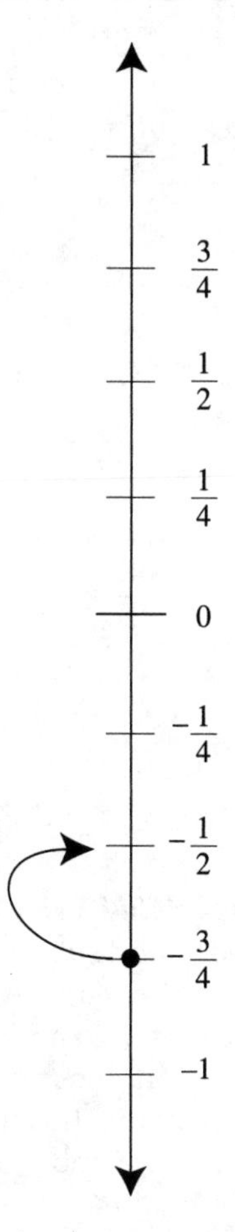

Try subtracting the fractions on the next page:

Quiz #6

1. $\frac{5}{6} - \frac{1}{2} =$

2. $-\frac{2}{3} - \frac{1}{8} =$

3. $\frac{3}{4} - \frac{3}{4} =$

4. $\frac{3}{2} - \left(-\frac{1}{4}\right) =$

5. $-\frac{13}{10} - \left(-\frac{4}{5}\right) =$

6. $\frac{50}{3} - \frac{5}{8} =$

7. $\frac{54}{7} - \frac{3}{14} =$

8. $\frac{3}{4} - \frac{1}{6} =$

QUICK WAYS TO ADD AND SUBTRACT FRACTIONS

So far, you have been adding and subtracting fractions by finding a common denominator. An easy way to find a common denominator for two fractions is to multiply their denominators. You can also make a common denominator within the fraction problem itself, instead of going through the process of multiplying each fraction. And remember, numbers can have many different common denominators.

Look at $\frac{1}{2} + \frac{2}{5}$ again. You can multiply 5 (the denominator of $\frac{2}{5}$) by 1 (the numerator of $\frac{1}{2}$) and get 5.

"Oh my gosh," you might be saying to yourself, "this is eerily similar to how we talked about comparing fractions not too many pages back." (Approximately how many pages back?)

$$\overset{5}{}\frac{1}{2} \nwarrow\!+ \frac{2}{5}$$

Then you can multiply 2 (the denominator of $\frac{1}{2}$) by 2 (the numerator of $\frac{2}{5}$) and get 4.

$$\overset{5}{\frac{1}{2}} \times \overset{4}{\frac{2}{5}}$$

Now that you have multiplied the numerators and the denominators of the other fractions, *you have the same numerators as if you had multiplied* $\frac{1}{2}$ *by* $\frac{5}{5}$, *or* $\frac{2}{5}$ *by* $\frac{2}{2}$.

You can now multiply the denominators together, to get 10.

$$\overset{5}{\frac{\not{1}}{2}} + \overset{4}{\frac{\not{2}}{5}} = \frac{\quad}{10}$$

Now, add your new numerators together, and put them on top of your new denominator.

$$\overset{5}{\frac{\not{1}}{2}} + \overset{4}{\frac{\not{2}}{5}} = \frac{9}{10}$$

This will save you lots of time working out separate computations for one question. It works just the same for subtraction except that you *subtract* the new numerators.

$$\overset{5}{\frac{1}{2}} - \overset{4}{\frac{2}{5}} = \frac{1}{10}$$

Try it with the problems on the next page:

Quiz #7

1. $\frac{2}{3}+\frac{5}{7}=$

2. $\frac{4}{5}-\frac{1}{3}=$

3. $\frac{2}{7}+\left(-\frac{2}{5}\right)=$

4. $\frac{3}{8}-\frac{1}{5}=$

5. $-\frac{7}{8}-\left(-\frac{1}{3}\right)=$

6. $-\frac{3}{5}+\frac{2}{3}=$

7. $\frac{12}{13}-\left(-\frac{3}{7}\right)=$

8. $\frac{1}{9}+\frac{1}{10}=$

Multiplying Fractions

Multiplying fractions is actually simpler than adding and subtracting them. To find the answer to a question like, "If you have 2 bags of laundry that have in them 3 shirts each, how many shirts do you have?" you multiply. The way you recognize when a question is asking you to multiply is when it says "of" something. This is true when the numbers involved are integers, and when the numbers involved are fractions. If I have $\frac{1}{2}$ of a group of 4 candy bars, I have 2 candy bars, right? To figure $\frac{1}{2}$ of 4, multiply $\frac{1}{2}$ by $\frac{4}{1}$. When you multiply fractions, you multiply the numerators and the denominators straight across.

$$\frac{1}{2}\times\frac{4}{1}=2$$

Multiplying fractions is something you've already been doing, though you might not have called it that. When you form a common denominator you *multiply* the fraction by 1, in the form of $\frac{2}{2}$, $\frac{5}{5}$, or some other form of 1.

A question like, "If you have $\frac{1}{5}$ of a bag of laundry that is $\frac{2}{3}$ full, what fractional part of the bag of laundry do you have?" works the same way. The "of" in the question is asking you to multiply.

$$\frac{1}{5} \times \frac{2}{3} = \frac{2}{15}$$

You would have $\frac{2}{15}$ of a bag of laundry if you had $\frac{1}{5}$ of a bag that was $\frac{2}{3}$ full. When multiplying fractions smaller than 1, your product is smaller than the fractions you are multiplying with, because the fractions smaller than 1 mean only a part of the whole.

If you have $\frac{1}{2}$ an apple, and then take $\frac{1}{3}$ "of" that $\frac{1}{2}$, you would end up with only a tiny piece.

You multiply the numerators AND you multiply the denominators.

When you multiply fractions, you don't need to find a common denominator. Isn't that great?

Once you have a product, you reduce, just as you do with all fractions.

$$\frac{2}{3} \times \frac{3}{4} = \frac{6}{12} = \frac{1}{2}$$

When you have a fraction multiplied by an integer, how will you handle it? Remember how you could write an integer as a fraction.

$$5 = \frac{5}{1}$$

$$\frac{2}{3} \times 5 = \frac{2}{3} \times \frac{5}{1} = \frac{10}{3}$$

You could now write $\frac{10}{3}$ as a mixed number: $3\frac{1}{3}$.

Negative Fractions and Multiplying

When you multiply negative fractions you treat them the same way you treat negative whole numbers. Just as when you multiply a negative whole number times a positive whole number, your product in this kind of multiplication will be a negative number. And when you multiply a negative fraction times another negative fraction, your product will be positive.

Look at it this way: you owe 4 people $\frac{1}{2}$ a dollar each. So you have

$$\frac{4}{1} \times -\frac{1}{2}$$

And it's as though you add $-\frac{1}{2} + -\frac{1}{2} + -\frac{1}{2} + -\frac{1}{2}$

$$\frac{4}{1} \times -\frac{1}{2} = -\frac{4}{2}$$

Which becomes $-\frac{2}{1}$ or -2. You have $-\$2$.

Look at a problem in which a negative is multiplied by a negative.

$$-\frac{4}{5} \times -\frac{3}{7} = \frac{12}{35}$$

This negative positive stuff should look pretty familiar now. If it doesn't, take a look back on page 6.

Try these:

Quiz #8

1. $\frac{4}{5} \times \frac{2}{3} =$

2. $\frac{3}{6} \times \frac{8}{9} =$

3. $\frac{3}{4} \times -\frac{2}{7} =$

4. $\frac{1}{6} \times \frac{4}{9} =$

5. $-\frac{4}{5} \times -\frac{2}{3} =$

6. $\frac{1}{7} \times 4 =$

7. $\frac{2}{3} \times 3 =$

8. $-\frac{5}{9} \times \frac{2}{3} =$

QUICK WAYS TO MULTIPLY FRACTIONS

One way to get through fraction multiplication problems a little bit faster is to use a process called **cancellation**. When the numerator of one fraction and the denominator of another fraction have a common factor, you can reduce those numbers by that common factor. So you are "canceling" the original numbers and putting in smaller easier to work with numbers.

It's another diagonal operation. But it does not work with adding or subtracting, okay? Just multiplying.

$$\frac{2}{3} \times \frac{7}{8}$$

Since both 2 and 8 have the common factor 2, you can divide them both by 2 to get 1 and 4.

$$\frac{\overset{1}{\cancel{2}}}{3} \times \frac{7}{\underset{4}{\cancel{8}}} = \frac{7}{12}$$

After you go through this cancellation process, you'll have easier numbers to multiply, plus you get to skip the step of reducing the product.

You can do cancellation with both sets of numerators and denominators.

$$\frac{\overset{1}{\cancel{4}}}{\underset{3}{\cancel{9}}} \times \frac{\overset{1}{\cancel{3}}}{\underset{4}{\cancel{16}}} = \frac{1}{12}$$

You can do this whether the fractions are positive, negative, or both. Leave the signs in the problem as they are, and cancel around them. Look at this next problem to see what this means.

$$-\frac{\overset{2}{\cancel{6}}}{\underset{1}{\cancel{7}}} \times \frac{\overset{2}{\cancel{14}}}{\underset{1}{\cancel{3}}} = -\frac{4}{1} = -4$$

Try these:

Quiz #9

1. $\frac{5}{9} \times \frac{3}{4} =$

2. $\frac{8}{9} \times 6 =$

3. $\frac{5}{8} \times -\frac{16}{25} =$

4. $\frac{16}{24} \times \frac{8}{21} =$

5. $-\frac{3}{4} \times -\frac{2}{5} =$

6. $\frac{5}{13} \times \frac{4}{15} =$

7. $-7 \times \frac{4}{15} =$

8. $-\frac{5}{6} \times -\frac{12}{35} =$

DIVIDING

You've been reducing fractions by 1 in the guise of $\frac{2}{2}$, $\frac{5}{5}$, or some other form. But reducing is not the same as dividing, and dividing fractions is a little bit different from multiplying them.

To divide by a fraction, flip the fraction you're dividing by, and then multiply. A lot of people like to memorize this by saying, "When dividing don't ask why, just flip it over and multiply." But don't worry, you can ask why if you want to, just wait a few pages. What you are really doing is multiplying by that fraction's **reciprocal**.

For instance, the reciprocal of $\frac{3}{4}$ is $\frac{4}{3}$. The reciprocal of $\frac{1}{5}$ is $\frac{5}{1}$ or 5. The reciprocal of 4 is $\frac{1}{4}$. The reciprocal of a number is that number **inverted**, or turned upside down.

Fractions equal to 0, like $\frac{0}{4}$, have no reciprocals.

To divide a fraction by another fraction, take the fraction you are dividing into, and multiply it by the flip of the fraction you are dividing by. It looks like this:

$$\frac{1}{2} \div \frac{3}{4} \text{ becomes } \frac{1}{2} \times \frac{4}{3}$$

Now that it's a multiplication problem, treat it just like any other multiplication fraction problem in the world. With this problem, you can cancel and then multiply straight across.

$$\frac{1}{2} \times \frac{4}{3} = \frac{2}{3}$$

APPROXIMATE THIS:

How many regular-sized chocolate bars (the flat plain chocolate without almonds kind) will it take to cover up the whole side of a record album? How about a compact disc?

Sometimes a fraction division problem will look like stacked-up fractions. Since you know that the fraction bar is just another way to show a division sign, just divide away.

$$\frac{\frac{2}{3}}{\frac{5}{7}} = \frac{2}{3} \times \frac{7}{5} = \frac{14}{15}$$

You can treat division problems that include both fractions and integers the same way.

$$\frac{\frac{4}{5}}{5} = \frac{4}{5} \times \frac{1}{5} = \frac{4}{25}$$

$$\frac{6}{\frac{1}{3}} = 6 \times \frac{3}{1} = 18$$

Remember, once you have a division problem set up as a multiplication problem, you can cancel to multiply more quickly.

BUT NEVER NEVER CANCEL THE FRACTIONS BEFORE INVERTING THE DIVISOR!

Okay?

IF YOU WANT TO KNOW WHY FLIPPING THE FRACTION WORKS, READ THIS

Think about dividing regular numbers for a minute. Say you wanted to split a group of 60 people into 2 teams. You would divide 60 by 2. And it could look like $60 \div 2$, $2\overline{)60}$, or $\frac{60}{2}$. Take a look at that last one. What would you have if you multiplied 60 by the reciprocal of 2?

$$60 \times \frac{1}{2} = \frac{60}{2} = 30$$

So all you've been doing this whole time to divide is multiplying your dividend by the reciprocal of your divisor. This goes back to what you've been seeing all along, that division is the inverse operation of multiplication. Everything in math is connected.

What do you think happens when you divide negative fractions?

It works like other division. When you divide a negative fraction by a positive fraction, or a positive fraction by a negative fraction, the result will be negative. Think of it this way. If you split up a debt of $\frac{1}{2}$ a dollar among 3 people, each person will get a piece of the debt, not any positive money.

$$-\frac{1}{2} \div \frac{3}{1}$$

$$-\frac{1}{2} \times \frac{1}{3} = -\frac{1}{6}$$

Each person now has a debt of $\frac{1}{6}$ of a dollar.

When you divide a negative fraction by another negative fraction, the result will be positive.

$$-\frac{1}{2} \div -\frac{3}{4}$$

$$-\frac{1}{2} \times -\frac{4}{3} = \frac{4}{6} = \frac{2}{3}$$

Try these division problems:

Quiz #10

1. $\dfrac{\frac{3}{4}}{\frac{5}{6}} =$

2. $4 \div \frac{1}{2} =$

3. $-\frac{1}{8} \div -7 =$

4. $\dfrac{\frac{4}{5}}{-\frac{6}{7}} =$

5. $\frac{2}{3} \div -\frac{1}{4} =$

6. $\dfrac{-\frac{3}{2}}{-\frac{9}{8}} =$

7. $1\frac{1}{3} \div \frac{2}{5} =$

8. $-3\frac{1}{3} \div \frac{4}{3} =$

USING FRACTIONS IN ALL KINDS OF PROBLEMS

Now that you are a fraction superstar, you can do any problem that includes fractions.

Quiz #11

1. Twenty pencils are removed from a box of 100 pencils. What fraction of the pencils remain in the box? (Sound familiar?)
2. In a bookshelf containing exactly 174 books, there are 15 German books, 25 French books, and 37 Chinese books. The remainder of the books are unclassified. What fractional part of the books is neither German nor French?
3. A news magazine contains 5 articles about finance and 3 articles about popular culture. The remaining $\frac{1}{3}$ of the articles in the magazine are about educational choices. What fraction of the articles are about finance?
4. If $\frac{1}{4}$ the recommended daily requirement of vitamin P is 2 ounces, what is the total recommended daily requirement of vitamin P?
5. A company wants to send out a mailer. A mailer can be sent only if it has an envelope and a stamp. For the number of mailers printed, the company has only $\frac{1}{2}$ the number of envelopes needed. Of those mailers for which there are envelopes, the company has only $\frac{1}{3}$ the number of stamps needed. What fractional part of the printed mailing can the company send out with the supplies it has?
6. How many $\frac{1}{2}$-ounce pencils are there in a box of pencils weighing 154 ounces? (Assume that only the pencils figure into the weight of the box of pencils.)

7. To feed a group of hungry students, a cafeteria must produce 200 pounds of glop. Their stoves, however, only produce $\frac{1}{5}$ of a pound of glop with each run. How many times must the cafeteria run the ovens to produce the 200 pounds of glop?

8. Brad owes a lot of money. He owes his mother $\frac{1}{4}$ of his future earnings, his dog $\frac{1}{16}$ of his future earnings, and his school $\frac{3}{32}$ of his future earnings. If you could put your answer in terms of his negative amount of money right now, what fractional part of his future earnings does Brad have? (You are just adding negative fractions.)

DECIMALS

Think back to the box of pencils for a moment. You know you had 100 pencils in the box, and you were in the process of giving someone 20 of that original 100, or $\frac{20}{100}$, or $\frac{1}{5}$. Here is yet another way to express the same thing.

$$0.2 = \frac{20}{100} = \frac{1}{5}$$

Decimals are just one more way to express fractional parts. The thing to understand about decimals is that they are fractions over powers of ten.

"What?" you say.

Well, you know what powers of ten are from when you looked at exponents. If you don't remember them too well, take a look back on page 48. 10^2 is a power of ten, and it is equal to 10. 10^2 is also a power of 10, it's equal to 100. Remember? Is it all coming back to you now? Here's how it works with decimals. Earlier, digits places were discussed. We talked about digits places to the left of the units digit. Well guess what, there are digits places to the right of the units digit, too—they are called decimal places. The point separating the decimal places from the units or ones digit is called the **decimal point**.

0.57

When there is a ∅ or no number to the left side of the decimal point it means your decimal has a value less than 1, like the decimal above. To read it aloud you say, "point five seven." The way decimals work is that each decimal place represents one of those fractions over powers of ten we were talking about. The place just to the right of the decimal point is the **tenths place**.

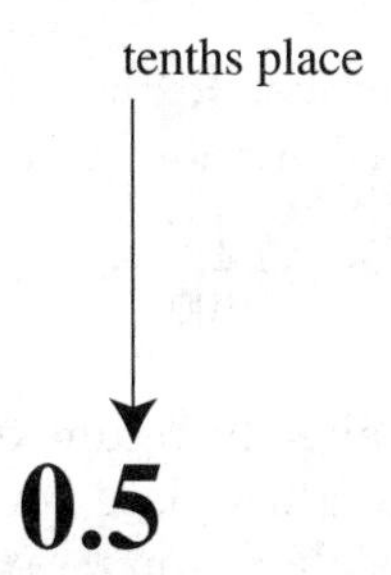

So the decimal 0.5 could also be written as $\frac{5}{10}$, 0.3 would be $\frac{3}{10}$, and 0.1 would be $\frac{1}{10}$.

The place to the right of the tenths place is the **hundredths place**.

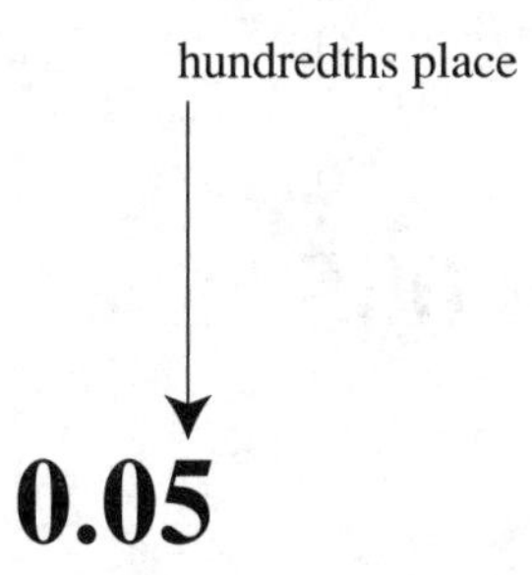

The decimal 0.05 can also be written as $\frac{5}{100}$. And 0.15 can be written as $\frac{15}{100}$. Going back to the first example, fractionally, 0.57 would be written $\frac{57}{100}$. You have 57 out of a possible 100. When it is written as a decimal people like to write in the 0 next to the .57 even though they would never think of doing that with a fraction.

How do you see 0.57? Probably the way you view most decimals in the world, as money. It is 57 cents. You have 57 out of a possible 100 cents, since the U.S. monetary system is based on a power of 10 (100 cents to the dollar).

34.12

When there is a number to the left of the decimal point, it is like having a mixed number in a fraction. 34.12 would be written fractionally as $34\frac{12}{100}$ or $34\frac{3}{25}$. Or 34 dollars and 12 cents.

You can probably already figure out what the place just to the right of the hundredths place is called: the **thousandths place**, though you probably won't say any of these decimal places aloud because you would end up spitting all over everything.

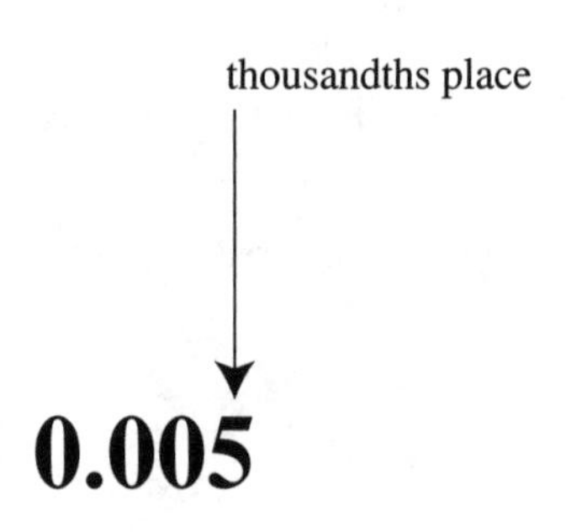

You can write 0.003 as $\frac{3}{1,000}$, 0.035 as $\frac{35}{1,000}$, and 0.123 as $\frac{123}{1,000}$.

You know that money is always written as decimals when there is any change that doesn't make up a whole dollar. This gives you an easy way to approximate with decimals. For instance, if you were trying to add 0.024 and 0.341 you might think to yourself, "Well if I forget about the ends it's just like adding 2 cents to 34 cents so I am going to have around 36 cents or 0.36."

Why is it that you can drop the ends sometimes?

The digit 4 at the end of 0.024 represents $\frac{4}{1,000}$, which is a very tiny number. It's as though you cut an apple into 1,000 equal pieces, which would be tiny, and took 4 of them. Now you won't be able to completely ignore these numbers always, but it is always a help to approximate, as you know. Approximating takes on a form with decimals called **rounding**.

0.024 gets rounded to 0.02

When you are rounding numbers, you are shortening them according to their decimal places. These are the rules of rounding:

When rounding a number, if the digit to the right of the place you are rounding to is greater than or equal to 5, you take it away and make the digit to the left 1 greater.

0.035 rounded to the nearest *hundredth* becomes 0.04
0.167 rounded to the nearest *hundredth* becomes 0.17
125 rounded to the nearest *tens place* becomes 130

Look at that last one. Sometimes you will be asked to round nondecimal numbers. Rounding appears in the form of a request to round a number to a particular decimal place.

The other rule of rounding is this: if the digit to the right of the place you are rounding to is less than 5, you throw out the numbers to the right and leave the number to the left as it is.

0.032 rounded to the nearest *hundredth* becomes 0.03
1.541 rounded to the nearest *hundredth* becomes 1.54
323 rounded to the nearest *tens place* becomes 320

Sometimes you will be asked to round off several places. You will have a number that goes to the thousandths place, and you will be asked to round it to the nearest tenth. Just find the digit to the direct right of the place to which you are rounding.

Someone says, "Round 78.6345 to the nearest tenth."

78.6345

You find the place next to the tenths place, the hundredths place, and the number there is 3. Since 3 is less than 5, you drop it, and leave the 6 in the tenths place as is. Don't worry about the numbers to the right of the 3, if you are rounding the number to the nearest tenth, look only to the direct right of the tenths place.

78.6

Now you have rounded it to the nearest tenth.

Try a few more and pay CLOSE attention to the place you need to round them.

Quiz #12

1. Round 0.45 to the nearest tenth.
2. Round 145.625 to the nearest tens place.
3. Round 0.04050607 to the nearest hundredth.
4. Round 10.020307 to the nearest tenth.
5. Round 53.6274 to the nearest hundredths place.
6. Round 13.573 to the nearest tenth.
7. Round 32.435 to the nearest hundredth.
8. Round 0.1234 to the nearest tenth.

APPROXIMATE THIS:

How many marshmallows (regular size) would it take to fill the room you are in?

Another way to look at decimals is to think of them as the results of division when without the term "remainder."

Here's how it works:

As you know, $\frac{5}{10}$ is $\frac{1}{2}$. And $\frac{5}{10}$ is also 0.5. You know that because you know $\frac{5}{10}$ has 10 as its denominator. Here is another way to see it.

$$10\overline{)5}$$

When you are dividing 5 by 10, you know 10 won't fit into 5. What you can do here is divide the 10 into 50. How? By adding on a 0 to the 5, a 0 that will be to the right of 5's decimal point.

$$10\overline{)5.0}$$

Now you can divide 10 into 50. You know 10 goes into 50, 5 times, so you write in the 5 above the 50. Since you are dividing into the 50, you write the 5 of your quotient above the 0 in 50, which means to the right of the decimal point.

$$\begin{array}{r} .5 \\ 10\overline{)5.0} \end{array}$$

There is your result, 0.5.

Try dividing these numbers so you get an answer in decimal form.

Quiz #13

1. $45 \div 6 =$
2. $30 \div 4 =$
3. $5 \div 25 =$
4. $17 \div 5 =$
5. $21 \div 6 =$
6. $9 \div 100 =$
7. $5 \div 4 =$
8. $345 \div 20 =$

Sometimes when you divide this way the answer will seem to go on forever. That is because actually, the answer *does* go on forever. When the answer does the same pattern forever, it is called a **repeating decimal**.

$$3\overline{)10}$$

becomes

$$\begin{array}{r} 3.3333333333333333333 \\ 3\overline{)10.0000000000000000000} \\ \underline{9} \\ 10 \\ \underline{9} \end{array}$$

The way to write this so you don't have to take up the whole page is 3.33. Of course it is probably a whole lot easier to write it as $3\frac{1}{3}$.

Another thing that can happen is that you divide and the decimal goes on forever and doesn't repeat itself in any pattern. When a decimal goes on forever and doesn't repeat, it is called a **nonrepeating decimal** (creative, huh?). It is also called—this is scary—an **irrational number**.

An example of a nonrepeating decimal is π, or "pi," which you'll hear more about in a while. If you measure a circle around its edge, and then divide that by its across-the-face measure, you get a nonrepeating decimal.

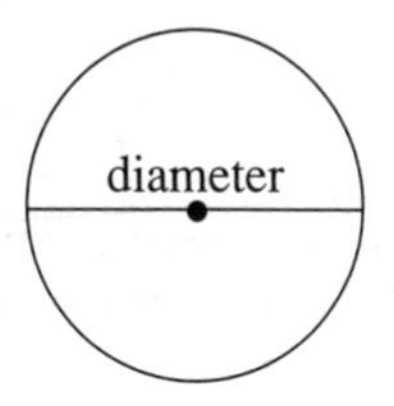

$$\pi = 3.14159265358979323846\ldots$$

We could go on, but most people round π off to 3.14.

To write nonrepeating decimals in a shorter form, you just round them off when you get too tired to write any more numbers. Mostly only scientists and mathematicians write them on and on forever, looking for patterns. In day-to-day arithmetic, rounding off will be called for a lot. Especially if the United States stops issuing the penny, which some people say is a likely next step in our monetary history. Hold on to those pennies, maybe they will be collectors' items someday.

Adding Decimals

You're saying, "What's so great about decimals?"

One of the great things about decimals is that since they are fractions with powers of ten, you don't have to do anything fancy when you add them. Think about it; their denominators are so similar. All you have to do is make sure that you line the decimal places up evenly so you are adding tenths to tenths, hundredths to hundredths, and all that sort of thing.

Just like with fractions, you wouldn't want to combine fractions with different denominators without thinking about it. And adding decimals will be no problem for you.

Just think about money. A dollar is 100 pennies, so a penny is $\frac{1}{100}$ of a dollar or 0.01. So you can think about adding money if you ever get confused as to how things should add up.

To add decimals, line them up along the decimal point and add as you would integers.

$$\begin{array}{r} 0.45 \\ \underline{+0.1} \end{array}$$

If, as in the problem above, the numbers don't look like they're the same length—don't worry about it. Just write in a 0 for any missing place and add.

$$\begin{array}{r} 0.45 \\ \underline{+0.10} \\ 0.55 \end{array}$$

Add a few more. There are some negatives in here, too, and you handle them the same way you would handle any negative numbers in addition. If you have any doubts about which direction to go, let the number line guide you.

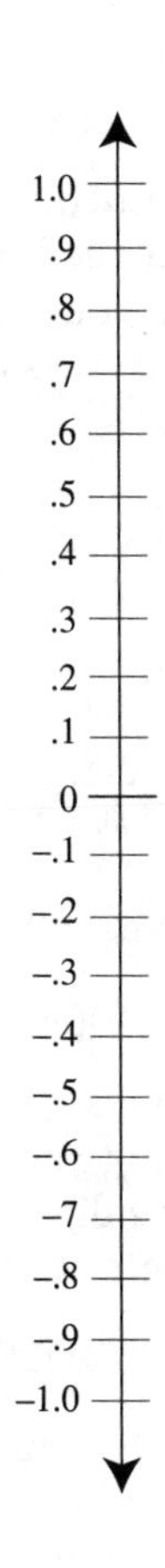

Quiz #14

1. $0.01 + 0.3 =$
2. $-1.523 + 6.7 =$
3. $0.001 + 0.01 =$
4. $32.32 + -11.1 =$
5. $60.12 + 0.003 =$
6. $3.123 + 5 =$
7. $-0.02 + -1.75 =$
8. $0.5 + 0.5 =$

SUBTRACTING DECIMALS

Subtracting decimals works much the same way as adding them. Think of fractions again: you need to add and subtract common denominators, so make sure your decimal places are lined up. When one looks shorter, just throw a few 0s onto the end to make it line up.

$$0.01 - 0.002$$

becomes

$$\begin{array}{r} 0.010 \\ \underline{-0.002} \\ 0.008 \end{array}$$

You know from working with fractions that numbers can look different and have the same value, like $1 = \frac{2}{2} = \frac{7}{7}$. The same thing is true of decimals. You can add 0s to the end of a decimal without changing the value of the number, since you are not moving any of the digits with respect to the decimal point. Adding 0s to a whole number <u>will</u> change it.

12 to 120

In that case, you move the digits farther to the left of the decimal point, which changes the value of the number. This is just something to keep in mind. If you ever get tempted to start throwing 0s onto the ends of numbers to the left of the decimal point, resist.

Try a few subtraction questions. When you are subtracting a bigger number from a smaller number remember to put the bigger number first and then just tag the negative sign onto your answer when you're done.

Quiz #15

1. $0.45 - 0.005 =$
2. $-4.5 - 3.2 =$
3. $0.05 - 1.5 =$

4. $3.45 - 0.1 =$
5. $0.005 - 0.0005 =$
6. $25.306 - 13.002 =$
7. $0.002 - 3.15 =$
8. $23.5 - 23 =$

Doing these exercises—possibly not the most exciting experience of your life—can help you to figure out whether one decimal is bigger than another. To compare decimals, look at the decimal places, starting with the tenths place. Try this on the example below:

$$0.0101 \text{ or } 0.0010$$

Comparing these numbers, start with the tenths place. Since both numbers have $\frac{0}{10}$, the tenths places are the same and you have to look at the hundredths place. The decimal on the left has $\frac{1}{100}$, and the decimal on the right has $\frac{0}{100}$, so the decimal on the left is bigger.

Another way to compare decimals is to convert them to fractions with the same denominator.

$$\frac{101}{10{,}000} \text{ or } \frac{10}{10{,}000}$$

If you cut something good up into 10,000 pieces, would you rather have 101 of those 10,000 pieces, or 10 of those 10,000 pieces? 101 would give you more.

And what about $\frac{10}{1{,}000}$ anyway? Can you look at this fraction in another way? How about $\frac{1}{100}$? Fractions in which both the numerator and the denominator end in 0 mean you can drop the 0s and it will be fine.

EXTREMELY HELPFUL THINGS TO THINK ABOUT

Do you remember that one of the easy ways to multiply a number by 10 is to add a 0 to the end of it? Well what you are really doing there is moving the decimal point to the right.

$$5 \times 10 = 50$$

This makes sense: take a minute to think about it. When you multiply by 10, you are increasing the number by a power of 10, which is one decimal place. This works when you multiply numbers by other powers of 10 as well. Remember, powers of ten are the results of 10 being raised to a power, an exponent, like on page 48. If you multiply 3 by 100, you move the decimal place in 3 two spaces to the right.

$$3 \times 100 = 300$$

It works when dividing by 10 or powers of 10 as well. But when you are dividing, you move the decimal point to the left.

$$50 \div 10 = 5$$

Think of this as a fraction division. $\frac{50}{10}$ becomes $\frac{5}{1}$ or 5. And all you are doing is taking that 50 and moving the decimal place once to the left. This works for all powers of 10 on all kinds of numbers (as long as they are in decimal form rather than fraction bar form).

Why is it that you can move the decimal place once to the left when you don't even see the decimal place? Because the decimal place is always there, even if it isn't written out. Think of money again: you can write $20 or you can write the same amount $20.00. The decimal is there, it is just not visible.

APPROXIMATE THIS:
How many golf balls would it take to fill up an elevator (the regular-sized office building kind)?

Try your hand at these powers of ten:

Quiz #16

1. $35 \times 100 =$
2. $10 \times 10 =$
3. $5{,}800 \div 100 =$
4. $3.45 \times 10 =$
5. $0.098 \times 1{,}000 =$
6. $123 \div 10 =$
7. $65.78 \div 100 =$
8. $100{,}000 \div 100 =$

Multiplying Decimals

Since you know how to multiply regular numbers, you are eminently qualified to begin multiplication of decimals. If you had 4 stacks of 50 cents, how much money would you have?

$$4 \times 0.5 =$$

What you do is first, multiply the numbers as though they were integers.

$$4 \times 5 = 20$$

Then, count how many decimal places, counting from right to left, you had in your *first* set up. You had one place, the tenths, so you make sure you put one decimal place in your product, starting from the right.

$$4 \times .5 = 2.0$$

You would have $2.00 if you had 4 stacks of $0.50 each. To figure out where the decimal point goes, just count up the number of decimal places in the numbers you are multiplying and put them in your product, starting from the right.

$$3.5 \times 1.7 =$$

Remember, first, multiply the numbers as though they had no decimals at all.

$$\begin{array}{r} 35 \\ \times\ 17 \\ \hline 245 \\ 350 \\ \hline 595 \end{array}$$

Then, count how many decimal places you have in the numbers you are multiplying, and add them together. You have one place in 3.5, and one place in 1.7. Together, that gives a total of two places. Put in the decimal places starting at the right.

$$5.95$$

Another way to make sure you have put the right number of decimal places in your answer is to approximate. If you rounded 3.5 and 1.7, what would you have? 4×2. So you know your product will be somewhere near 8. Not 80 or 800.

Try multiplying a few more. Approximate the answers first, and write down your approximation. And remember, it never hurts to be thinking about the distributive property, page 31.

Quiz #17

1. $0.66 \times 4.2 =$
2. $1.53 \times 32 =$
3. $0.03 \times 4.001 =$
4. $12.1 \times 100.1 =$
5. $3 \times 5.5 =$
6. $0.0002 \times 4 =$
7. $100 \times 5.63 =$
8. $0.2 \times 0.2 =$

Great. Now you know how to handle tons of different things that come up with decimals. By the way, did you use the distributive property on number 4? 12.1 times 100 just moves the decimal over two to the right to get 1,210. You add this to 12.1 times the remaining .1. Just to remind you, multiplying by .1 is the same as multiplying by $\frac{1}{10}$. Which means it is the same as dividing by 10, which means you can just move that old decimal place to the left. You get 1.21. Add the 1,210 and the 1.21 together and you have 1,211.21, your answer.

And if you were multiplying by .01 you would move it two places to left, and if you were multiplying by .001, well . . . it's all connected.

APPROXIMATE THIS:

How many pennies are in $1 million?

Decimal Division

One of the clearest ways to look at decimal division is to keep this guideline in your head: the divisor should always be in the form of an integer.

$$.3\overline{).45}$$

Since you just read that whole thing about "keep it in the form of an integer," you know you don't want to look at it as .3, but rather 3, and you may be thinking, "How do I do that?" Just move your decimal place to the right until you have a whole number. In this case, that means you have to move it to the right only once. Once you have transformed your divisor, give the same treatment to your dividend.

$$3.\overline{)4.5}$$

Write the decimal point above the place it sits in the dividend. Then divide as you would normally.

$$3\overline{)4.5}^{\,1.5}$$

To check your work, you can multiply your quotient by your divisor and see if you get your dividend. This should seem familiar; if it doesn't, check back on page 41. And you should always practice approximating to see if your answer looks about right. Are there about 1.5 groups of .3 each in .45? Or how many groups of 30 cents are there in 45 cents? About $1\frac{1}{2}$.

Try a bunch:

Quiz #18

1. $35 \div .5 =$
2. $6.72 \div .42 =$
3. $6.5 \div 1.3 =$
4. $55 \div .11 =$
5. $0.58 \div 2 =$
6. $77.7 \div .001 =$
7. $.34 \div .17 =$
8. $500 \div .001 =$

Look at that last question. Would it be easier to use the .001 as a fraction and then multiply? Maybe. Just as multiplying by a decimal power of ten means moving the decimal point, dividing by a decimal power of ten means moving the decimal point, too. Which way did you move it here? To the right three times. When dividing by decimals smaller than 1, move the decimal point to the right. And always, always approximate. How many .001s, or $\frac{1}{10}$ of a cent, would fit in $500? Lots.

Check out some decimals used in word form, and test your formidable skills.

Quiz #19

1. Cynthia has 0.32 of a bowl of cherries, and she wants to split it equally among herself and her three friends. What part of this bowl of cherries does each person get?
2. Marilyn wants to divide $1.74 into piles of $0.03 each. How many piles is she going to end up making?
3. If you had 7.378 yards of dental floss and you wanted to make a dental floss string that was 673.5423 yards long, how many more yards of dental floss would you need? (Assume you don't need any extra to tie the two pieces together.)
4. If you had 17.5 jars of peanut butter and each jar was filled up to 0.54 ounces, how many ounces of peanut butter would you have?
5. Jane has 3,450.75 pounds of chocolate in her refrigerator. She decides to split it all between her two best friends. How many pounds of chocolate does each friend get, rounded off to the nearest tenth? (Remember, Jane does not get any of the chocolate.)
6. If you had no money to start with, and then someone gave you $110.95, and then you received $56.76 anonymously in the mail, and then you won $73.21 in a lottery (a lame lottery), and then you had to pay $43.80 for rent, how much money would you end up with, rounded to the nearest dollar?
7. How many stacks of coins worth exactly $0.35 each would you need to get $35.00?
8. Paul, Tom, and Geoff go to a restaurant for dinner. If the check equals $75.85, and they decide to leave $15.20 for the tip, how much does each guy owe, if they decide to split it equally (even though Paul is annoyed by this since he got the least expensive thing on the menu)?

PERCENTAGES

Now we get to **percentages**. And guess what. You know how you feel about decimals after looking at fractions? That whole why-they're-just-fractions-over-powers-of-10,-I'm-not-sure-why-they-seemed-so-terrible feeling? Well it's the same thing with percentages. The only difference is that percentages are an even smaller part of fractions. Percentages are fractions over 100. Percent, when you break it down, becomes per and cent. Per as in "there are 3 hats per every 5 children," which means you have $\frac{3}{5}$ of the hats you need. *Per* means for each. *Cent* means 100, cent is like *cent*ury or *cent*s as in 100 per (there it is again) dollar; so percent means for each 100.

50% means 50 for each 100. It can also look like .5, or $\frac{50}{100}$, or $\frac{1}{2}$ (just reduce the fraction).

100% means 100 for each 100 or $\frac{100}{100}$ which means 1. The whole. If you have 100 percent of something you have the whole thing.

CONVERTING FRACTIONS TO PERCENTAGES

If a fraction is already over 100, just drop the denominator and add the percent sign.

$$\frac{99}{100} = 99\%$$

If the fraction is over a power of 10, try to convert the denominator to 100 by multiplying by $\frac{10}{10}$, or $\frac{100}{100}$, or reducing to a power of 10 (just move the decimal point on the numerator). To review this section of common denominators, check back to page 68.

$$\frac{6}{10} = \frac{6}{10} \times \frac{10}{10} = \frac{60}{100} = 60\%$$

$$\frac{5}{1,000} = \frac{5 \div 10}{1,000 \div 10} = \frac{0.5}{100} = 0.5\%$$

And if you are starting from a decimal, translating to a percentage is even easier. The number .68 is 68%, because it is already over 100.

When you want to convert a fraction to a percentage and that fraction is not already over a power of 10, you need to convert the fraction to a decimal. This means dividing the numerator by the denominator, and then moving the decimal place two spaces to the right and adding a percent sign.

$$\frac{3}{8} = 8\overline{)3}$$

$$\begin{array}{r} 0.375 \\ 8\overline{)3.000} \\ \underline{24} \\ 60 \\ \underline{56} \\ 40 \\ \underline{40} \end{array}$$

Put the decimal point two places to the left, and there is your percent: 37.5%. You can also write it as $37\frac{1}{2}\%$. As a fraction, it is $\frac{375}{1,000}$ or back to your original friend, $\frac{3}{8}$. It's all the same number, isn't that amazing? See how to work it in the other direction now.

Converting Percentages to Standard Fractions

To convert percentages to fractions, put the percentage over 100 and drop the % sign. Then reduce.

$$60\% = \frac{60}{100} = \frac{3}{5}$$

Sometimes percentages have fractions in them, like $66\frac{2}{3}\%$. This means it is $66\frac{2}{3}$ over 100, or $66\frac{2}{3}$ divided by 100 (same thing, remember?). To work with this number more easily, try converting $66\frac{2}{3}$ from a mixed number. Once it is converted, you can divide by 100.

Converting from a mixed number, $66\frac{2}{3} = 3 \times 66 = 198 + 2 = 200$; $\frac{200}{3}$. Converting from mixed number review is on page 72.

$$66\frac{2}{3} \text{ becomes } \frac{200}{3}$$

$$\frac{200}{3} \div 100 = \frac{200}{3} \times \frac{1}{100} = \frac{2}{3}$$

Convert Percentages to Decimals

To convert percentages to decimals, move the decimal point two spaces to the left and drop the % sign.

$$75\% = 0.75 \text{ and } 2\% = .02$$

Fractions are decimals are percentages. For the following questions, if something is in percentage form, convert it to both a fraction (reduced) and a decimal. If it is in decimal form, convert it to a fraction (reduced) and a percentage. If it is in fraction form, you guessed it, convert it to a decimal and a percentage.

Quiz #20

1. $.005 =$
2. $59\% =$
3. $\frac{4}{5} =$
4. $0.89 =$
5. $\frac{1}{16} =$
6. $33\% =$
7. $1.42 =$
8. $\frac{1}{8} =$

The way percentages are usually used is the same way fractions and decimals are used: to represent part of a whole. "I have 400 oranges and I am going to give you 30% of them." Just as with other types of fractions, "of" means multiply. When calculating with percentages, you can choose which form you would prefer. So, to find out 30% of 400, some people like to use percentages as fractions over 100.

$$400 \times \frac{30}{100} = 120$$

Some people like to use decimals.

$$400 \times .3 = 120.0$$

Some people like to calculate by taking 10% of the number, in this case 400, and then multiplying the 10% by however many percent they are taking. It's easy to take 10% of something because you just move the decimal place once to the left, exactly as you do when you multiply by $\frac{10}{100}$, or $\frac{1}{10}$. Then, if you have 10% of something, you can easily find 30% by multiplying that 10% by 3. It's the distributive property again.

$$10\% \times 400 = 40 \text{ and } 40 \times 3 = 120$$

You get to decide which form you use. It is going to be helpful to you to be comfortable with all of these ways of looking at numbers, because you never know . . . sometimes fractions will be the easiest way, sometimes multiples of ten percent will seem easier than fractions. Just remember this, the more you understand what the numbers mean and what they stand for, the happier and more comfortable with math you are going to feel.

Try this percentage problem, and use percentages in the form of reduced fractions:

Henry owns 36 watches. He decides to give 25% of his watches to charity. How many watches does he give away?

The fraction form of 25% is $\frac{25}{100}$ which can be reduced to $\frac{1}{4}$.

$$\frac{1}{4} \times 36 = 9$$

He gave away 9 watches.

Now look at a problem using decimal forms of percentages.

Susan has a fleet of 125 trucks. A tax adviser convinces her to sell 8% of her trucks to cut down on her overhead costs. How many trucks will she sell?

Why use a decimal here? Well, why not? Either a fraction or a decimal is equally useful but just for the heck of it, use the decimal. And remember, "of" means multiply.

$$125 \times 0.08 = 10.00$$

She sells 10 trucks.

Now try a problem using multiples of 10 percent. It's really a way of using the distributive property again, isn't it?

There are 410 boxes of compact discs in a warehouse. An extremist organization plans to burn 70% of the boxes because the group finds the lyrics objectionable. How many boxes of compact discs are set to be burned?

You can figure quickly by moving the decimal point that 10% of 410 is 41. Then, multiply this 10% by 7, to get 70%.

$$41 \times 7 = 287$$

287 boxes of compact discs are set to be burned. Don't try it at home, any of you.

Practice these problems using all the different ways you just tried to work with percentages. Try the compact discs with decimals, the truck fleet with fractions. The more you practice, the stronger your math muscles will get.

And try a few of these:

Quiz #21

1. $51\% \times 200 =$
2. $20\% \times -17 =$
3. $33\frac{1}{3}\% \times 138 =$
4. $0.03\% \times 100 =$
5. $250\% \times -30 =$
6. $25\% \times 300 =$
7. $1\% \times 4{,}560 =$
8. $23\frac{1}{3}\% \times -1{,}000 =$

APPROXIMATE THIS:

Your brain is what percent of your overall body weight?

FANCY PERCENTAGES

There are some applications of percentages you may want to know, like interest and percent increase or decrease.

Interest is when a set percentage is figured out, usually on money, at specific times. For instance, if a bank gives you 5% interest on your savings account, calculated yearly, it means every year the bank will give you money, and the money given will be calculated by figuring 5% of the money you have in your savings account. If you have $100 in your savings account, at 5% yearly interest, you would earn $5 in one year. So interest earned per time period is the amount of money in the account times the percentage of interest.

Percentage increase or **decrease** is a way to calculate how much something has grown or shrunk. If you started this year with 200 books in your library, and you ended the year with 260 books in your library, you have increased your library by 60 books. This is a 30% increase in your library size.

Here is how to calculate it. You started with 200. The increase is 60. So you need to know, what percent of 200 is 60? The way to set this up for percentage change is to put the increase over the original number. $\frac{60}{200}$. To convert this into a percentage, multiply by 100 after you reduce and then add the % sign.

$$\frac{60 \div 2}{200 \div 2} = \frac{30}{100}$$

$$\frac{30}{100} \times 100 = 30$$

There has been a 30% increase in books.

Calculating percent decrease is the same. If last year you ate 100 apples, and this year you ate 25 apples, what was the percentage decrease in your apple consumption?

The change in the number of apples is $100 - 25$, or 75.

Change over original number is $\frac{75}{100}$.

$$\frac{75}{100} \times 100 = 75$$

There was a 75% decrease in your apple consumption.

So percent decrease can also be calculated as amount of change over original number, then multiply the result by 100 and add the % sign.

You now know a lot of what there is to know about the different kinds of fractions used in math.

GLOSSARY

fraction: A way of expressing the division of numbers by stacking one over the other.
numerator: The top part of a fraction.
denominator: The bottom part of a fraction.
improper fraction: A fraction in which the numerator is bigger than the denominator.
proper fraction: A fraction in which the denominator is bigger than the numerator.
common denominator: The same denominator of separate fractions.
lowest common denominator: The smallest possible common denominator of separate fractions.
mixed number: A number made up of both an integer and a fraction.
cancellation: Reducing the numerators or denominators of fractions that are being multiplied, by dividing by a common factor.
reciprocal: The inverse of a number.
decimal: A fraction expressed as a number to the right of the decimal point.
repeating decimal: A decimal that repeats a pattern of numbers to infinity.
nonrepeating decimal: A decimal that goes to infinity but does not repeat a pattern.
percent: Amount per 100.

Try these problems that combine all these excellent types of fractions.

Quiz #22

1. A box contains a number of items, among them exactly 10 hats, 3 books, and 7 statues. The remaining $\frac{1}{3}$ of the items are all blue T-shirts. What percent of the items in the box are books?

2. Five hundred people, and no others, gathered in front of the White House on a certain day to protest the government's immigration policies. Of these 500 people, 60% wore red arm bands showing solidarity with the immigrants. Of this 60%, 20% carried signs. How many people in front of the White House carried signs?

3. Forty lamps represent $\frac{1}{3}$ of the lamps in a museum, and lamps in the museum represent $\frac{1}{2}$ of the electrical appliances in the museum. How many electrical appliances are there in the museum?

4. If 0.5 of the 70 shoes in Emily's closet are shoes from France, and 0.01 of the shoes in Emily's closet are shoes from Italy, what percent of the shoes in Emily's closet are from neither France nor Italy?

5. $(50\% \times 400) + \left(\frac{1}{4} \times 400\right) =$

6. Jason has 300 jackets. Last year he had only 50 jackets (poor guy). His present wardrobe of 300 jackets represents what percent increase in his number of jackets?
7. You have exactly $1 million in the bank. Congratulations. If your money is earning $8\frac{1}{2}\%$ interest annually starting today—annually means per year—how much money will you have, including the interest you earn, exactly 1 year from now? (One million $= 1,000,000$.)
8. $(0.005 \times .003) + \left(\frac{1}{3} \times 24\right) + (20\% \times 1) =$
9. If Alice has 5 suitcases, and each suitcase is filled with 30 skateboards—they are big suitcases and she's storing the wheels somewhere else—and she decides to give you $\frac{1}{6}$ of her total number of skateboards, and she has no other skateboards in the world, how many skateboards will she give you?
10. You worried about math last year for 300,000 minutes. This year, you will worry about math for only 60 minutes. What percent decrease in minutes of math worry will you experience this year?

ANSWER KEY

Quiz #1

1. $\frac{2}{3}$
2. $\frac{1}{12}$
3. $\frac{5}{9}$
4. $\frac{1}{3}$
5. $\frac{1}{2}$
6. $\frac{13}{30}$
7. $\frac{1}{5}$
8. 1

Quiz #2

1. $\frac{9}{14}$
2. $\frac{28}{32} = \frac{7}{8}$
3. $\frac{229}{198}$
4. $\frac{33}{130}$
5. $\frac{51}{35}$
6. $\frac{29}{39}$
7. $\frac{83}{10}$
8. $\frac{13}{12}$

Quiz #3

1. $\frac{10}{3}$
2. $\frac{41}{3}$
3. $\frac{35}{6}$
4. $\frac{73}{7}$
5. $\frac{13}{2}$
6. $\frac{3}{2}$
7. $\frac{19}{5}$
8. $\frac{52}{7}$

Quiz #4

1. $1\frac{3}{4}$
2. $4\frac{1}{2}$
3. $3\frac{1}{4}$
4. $2\frac{7}{8}$
5. $6\frac{4}{5}$

6. $23\frac{1}{3}$

7. $2\frac{1}{7}$

8. $2\frac{2}{5}$

Quiz #5

1. $\frac{58}{35} = 1\frac{23}{35}$ Either way you want to write it is fine. Remember, they're equal.
2. $\frac{113}{12} = 9\frac{5}{12}$ You may have added the integers and then the fractions and gotten $8\frac{17}{12}$. Remember, $\frac{17}{12}$ is $1\frac{5}{12}$ so you can add this to 8 and get $9\frac{5}{12}$.
3. $\frac{49}{12} = 4\frac{1}{12}$
4. $\frac{21}{5} = 4\frac{1}{5}$
5. $\frac{15}{4} = 3\frac{3}{4}$
6. $\frac{281}{12} = 23\frac{5}{12}$
7. $\frac{32}{32} = 1$
8. 7

Quiz #6

1. $\frac{1}{3}$
2. $-\frac{19}{24}$
3. 0
4. $\frac{7}{4} = 1\frac{3}{4}$
5. $-\frac{1}{2}$
6. $\frac{385}{24} = 16\frac{1}{24}$
7. $\frac{105}{14} = 7\frac{1}{2}$
8. $\frac{7}{12}$

Quiz #7

1. $\frac{29}{21} = 1\frac{8}{21}$
2. $\frac{7}{15}$
3. $-\frac{4}{35}$
4. $\frac{7}{40}$
5. $-\frac{13}{24}$
6. $\frac{1}{15}$
7. $\frac{123}{91} = 1\frac{32}{91}$
8. $\frac{19}{90}$

QUIZ #8

1. $\frac{8}{15}$
2. $\frac{4}{9}$
3. $-\frac{3}{14}$
4. $\frac{2}{27}$
5. $\frac{8}{15}$
6. $\frac{4}{7}$
7. 2
8. $-\frac{10}{27}$

QUIZ #9

1. $\frac{5}{12}$
2. $\frac{16}{3}=5\frac{1}{3}$
3. $-\frac{2}{5}$
4. $\frac{16}{63}$
5. $\frac{3}{10}$
6. $\frac{4}{39}$
7. $-\frac{28}{15}=-1\frac{13}{15}$
8. $\frac{2}{7}$

QUIZ #10

1. $\frac{9}{10}$
2. 8
3. $\frac{1}{56}$
4. $-\frac{14}{15}$
5. $-\frac{8}{3}=-2\frac{2}{3}$
6. $\frac{4}{3}=1\frac{1}{3}$
7. $\frac{10}{3}=3\frac{1}{3}$
8. $-\frac{5}{2}=-2\frac{1}{2}$

QUIZ #11

1. $\frac{4}{5}$ remain.

   How It Works:
   This is a fraction subtraction problem. You have $\frac{100}{100}-\frac{20}{100}=\frac{80}{100}$. $\frac{80}{100}-\frac{40}{50}=\frac{4}{5}$. $\frac{4}{5}$ of the pencils remain in the box.

2. $\frac{67}{87}$ of the books are neither German nor French.

How It Works:

The parts of the fraction in this case are the types of books, for instance $\frac{15}{174}$ of the books are German. To find what part of the books is neither German nor French, add the numbers of German and French books together. This gives you 40 books. ALL the books that remain, not just the Chinese books, are neither German nor French.

$$\frac{174}{174} - \frac{40}{174} = \frac{134}{174}$$

$\frac{134}{174}$ of the books are neither German nor French. This can be reduced to $\frac{67}{87}$.

3. $\frac{5}{12}$ of the articles are about finance.

How It Works:

This problem uses fractions to give the idea of part over whole. There are 8 articles mentioned, (5 finance + 3 popular culture) with $\frac{1}{3}$ of the articles remaining.

$$\frac{3}{3} - \frac{1}{3} = \frac{2}{3}$$

That means the 8 articles are $\frac{2}{3}$ of the total articles. If 8 is $\frac{2}{3}$, how do you find $\frac{1}{3}$? By taking half of 8. $\frac{1}{2}$ of $\frac{2}{3}$ is really $\frac{1}{2} \times \frac{2}{3} = \frac{1}{3}$

You do the same thing with 8. $\frac{1}{2}$ of 8 is $\frac{1}{2} \times 8 = 4$

The $\frac{1}{3}$ of the articles, or 4, combine with the $\frac{2}{3}$ of the articles, or 8, to form $\frac{3}{3}$ of the articles, or the whole, 12. So you know that you have 12 articles in the magazine. What part of these are about finance? 5. The fractional part of the magazine that contains articles about finance is $\frac{5}{12}$.

4. 8 ounces is the recommended daily requirement.
   How It Works:
   To find the total daily requirement you need not $\frac{1}{4}$, but the whole, or $\frac{4}{4}$. To get $\frac{4}{4}$ from $\frac{1}{4}$ you multiply by 4, so you need to multiply the 2 ounces by 4 as well.
   $2 \times 4 = 8$. The recommended daily requirement of vitamin P is 8 ounces.

5. They can send $\frac{1}{6}$ of the printed mailers.
   How It Works:
   The company has $\frac{1}{2}$ of the envelopes needed. $\frac{1}{3}$ of these $\frac{1}{2}$ will have stamps. When the word "of" is used in a math problem, it means "multiplied by." For instance, if you had $\frac{1}{2}$ of a box of 8 sheets of paper, you would have $\frac{1}{2} \times 8$ or 4 sheets of paper. So this is only a fancy looking fraction multiplication problem. $\frac{1}{3} \times \frac{1}{2} = \frac{1}{6}$. The company can send out $\frac{1}{6}$ of its printed mailers.

6. There are 308 pencils in the box.
   How It Works:
   This is a division problem—how many pencils go into the box—and the only thing that makes it a little strange is that you are dividing by the fraction $\frac{1}{2}$.
   $154 \div \frac{1}{2}$ is really 154×2.
   There are 308 pencils in a box of pencils weighing 154 ounces.

7. They must run the ovens 1,000 times.
   How It Works:
   Again you have a division problem using a fraction because the question asks how many times $\frac{1}{5}$ goes into 200. $200 \div \frac{1}{5}$ or $200 \times 5 = 1{,}000$. The ovens must be run 1,000 times to produce 200 pounds of glop.

8. In negative terms, Brad has $-\frac{13}{32}$ of his future earnings right now.

How It Works:

You need to combine all of Brad's debt to figure out how much he owes. To combine, you are going to add. You can add in positive or negative, whichever you would rather.

$$-\frac{1}{4} + -\frac{1}{16} + -\frac{3}{32}$$

To multiply the denominators here would be a lot of numbers, you can instead see if the biggest denominator could be the common denominator. Since 32 is a multiple of both 4 and 16, you are in business.

$$\left(\frac{1}{4}\right)\left(\frac{8}{8}\right) = \frac{8}{32} \quad \left(\frac{1}{16}\right)\left(\frac{2}{2}\right) = \frac{2}{32}$$

$$\frac{8}{32} + \frac{2}{32} + \frac{3}{32} = \frac{13}{32}$$

Brad has $-\frac{13}{32}$ of his future earnings right now. By the way, it is highly unlikely you will ever have a negative fraction problem set up quite this way, but it is good to try and put things in real terms to help you understand them.

Quiz #12

1. 0.5
2. 150
3. 0.04
4. 10.0
5. 53.63
6. 13.6
7. 32.44
8. 0.1

Quiz #13

1. 7.5
2. 7.5
3. 0.2
4. 3.4
5. 3.5
6. 0.09
7. 1.25
8. 17.25

Quiz #14

1. 0.31
2. 5.177
3. 0.011
4. 21.22
5. 60.123
6. 8.123
7. −1.77
8. 1.0

Quiz #15

1. 0.445
2. −7.7
3. −1.45
4. 3.35
5. 0.0045
6. 12.304
7. −3.148
8. 0.5

Quiz #16

1. 3,500
2. 100
3. 58
4. 34.5
5. 98
6. 12.3
7. 0.6578
8. 1,000

Quiz #17

1. 2.772
2. 48.96
3. 0.12003
4. 1,211.21
5. 16.5
6. 0.0008
7. 563
8. 0.04

Quiz #18

1. 70
2. 16
3. 5
4. 500
5. 0.29
6. 77,700
7. 2
8. 500,000

Quiz #19

1. .08

   How It Works:
   She is splitting the cherries, so you need to divide. $.32 \div 4$ (herself and 3 friends).

   $4\overline{).32}$ Each person gets .08 of a bowl of cherries. You can multiply each person by his or her portion and check to see if you get the right amount.

2. 58

   How It Works:
   Again, you are dividing $1.74 into piles, so you need to divide.

   .03 $4\overline{)1.74}$ There will be 58 piles of $0.03 each. Multiply to check it.

3. 666.1643

   How It Works:
   You have 7.378 yards of dental floss, and you need to have 673.5423 yards. What is missing? You need to subtract and find out.

   $$\begin{array}{r} 673.5423 \\ -\ 7.3780 \\ \hline 666.1643 \end{array}$$

   He needs 666.1643 yards of floss to complete his rope. Add the two pieces up and see.

4. 9.45 ounces

   How It Works:
   You are combining all the peanut butter so you could add up each one or, to make it easier, you can multiply. $17.5 \times .54 = 9.45$ ounces of

peanut butter. Divide this up into 17.5 jars and see if you get .54 ounces each.

5. 1,725.4 pounds of chocolate per person

   How It Works:

   Well, to split it means you have to divide it between those 2 best friends.

   $$2\overline{)3,450.750}$$ = 1,725.375

   Rounding 1725.375 to the tenths place means looking right next to the tenths place, the hundredths place, and rounding that. Since it is 7, it's bigger than 5 so you add 1 to the 3 in the tenths place. 1725.4 pounds of chocolate per person.

6. $197

   How It Works:

   First you need to figure out how much money you have after all this, which means you need to add up all this money.

   $110.95 + 56.76 + 73.21 = 240.92$

   Then, you need to subtract the money you paid for your *How It Works:*rent.

   $240.92 - 43.80 = 197.12$

   Now, you need to round 197.12 to the nearest dollar, or to the units digit place. Since the number in the tenths place is smaller than 5, you can just drop it, and you have $197.00

7. 100

   How It Works:

   This is asking you "how many stacks" if you split $35.00 into $0.35 stacks. Since you are splitting it into stacks, it means divide.

   $$.35\overline{)35.00}$$ = 1 00

   You would end up with 100 stacks. Multiply it to make sure it works. (Really, we know you don't feel like it, but it is an excellent habit to get into, so why not start now?)

8. $30.35

   How It Works:

   They need to split the check and the tip. Since you probably don't feel like doing the division twice, why don't you combine the amounts and split the one big amount?

   $$3\overline{)91.05}$$ = 30.35

   Each of them owes $30.35. Did you approximate first?

Quiz #20

1. $\frac{5}{1,000}$ and 0.5%
2. $\frac{59}{100}$ and 0.59
3. 80% and 0.8
4. 89% and $\frac{89}{100}$
5. .0625 and 6.25% or $6\frac{1}{4}\%$
6. 0.33 and $\frac{33}{100}$
7. 142% and $1\frac{21}{50}$ or $\frac{71}{50}$
8. 0.125 and 12.5% or $12\frac{1}{2}\%$

Quiz #21

1. 102
2. $-\frac{17}{5}$ or −3.4
3. 46
4. 0.03
5. −75
6. 75
7. 45.6
8. $-\frac{700}{3}$ or −233.33 or $-233\frac{1}{3}$

Quiz #22

1. 10% of the items are books.
   How It Works:
   Since this problem talks to you about "the remaining $\frac{1}{3}$ of the items," you know that the 10 hats, 3 books, and 7 statues, represent the rest. You need all of the items to make up the whole; you need to figure out how many items are in the whole. Combine the items you have.

$$10 + 7 + 3 = 20$$

These 20 items represent $\frac{2}{3}$ of the items ((the whole) $\frac{3}{3} - \frac{1}{3}$ (T-shirts)). You want to find the whole, or $\frac{3}{3}$. One way to do this is to figure how many items are in $\frac{1}{3}$, and add it to your $\frac{2}{3}$, to get $\frac{3}{3}$. Divide $\frac{2}{3}$ in half, and you get $\frac{1}{3}$. So you divide 20 in half and get 10, which is the remaining $\frac{1}{3}$ of the stuff in the box. Add the 10 to 20 and you have 30 items in the box. What percent are books? Well, $\frac{3}{30}$ are books; convert it to a percentage.

$\frac{1}{10} = 10\%$

2. 60 people carried signs.
   How It Works:
   First you need to find out how many people had arm bands. 60% of the 500. You can write it $\frac{60}{100} \times 500 = 300$
   300 people had arm bands. Of these 300, remember they said of the 60%, 20% had signs.

   $\frac{20}{100} \times 300 = 60$

   60 people carried signs.

3. There are 240 electrical appliances in the museum.
   How It Works:
   You need to find all the lamps in the museum for a start. Since 40 lamps are $\frac{1}{3}$, to get $\frac{3}{3}$ or the whole, you multiply $\frac{1}{3}$ by 3. That means to find the whole of the lamps you multiply 40 by 3.

   $40 \times 3 = 120$

   There are 120 lamps, and lamps are $\frac{1}{2}$ of the appliances. To find the whole of appliances, or $\frac{2}{2}$, you would multiply by 2. So to find all the appliances,

   $120 \times 2 = 240$

   There are 240 electrical appliances in the museum.

4. 49% of her shoes are from neither France nor Italy.
   How It Works:
   Here they are using decimals and asking about percentages. Just transfer those decimals to percentages, and see what is left. $0.5 = 50\%$, and 0.01 is 1%, so you already know about 51% of her shoes, so the rest of her shoes, the ones not from France or Italy, are the remaining 49%. (Remember, 100% is the whole.)

5. 300
   How It Works:
   50% of

   $400 = \frac{50}{100} \times 400 = 200$

   $\frac{1}{4} \times 400 = 100$

   $200 + 100 = 300$

6. There has been a 500% increase in Jason's jacket wardrobe.
   How It Works:
   Percent increase is calculated by change over original number, and the result is then multiplied by 100. The change in his number of jackets is $300 - 50 = 250$.

   The number of jackets he started with is 50.

   $$\frac{250}{50} = 5$$
   $$5 \times 100 = 500$$

   Add the percentage sign and you are in business.

7. You will have $1,085,000 in one year.
   How It Works:

   Since you have $8\frac{1}{2}\%$ interest, it may be easier to put the interest as a decimal.

   0.085

   Because you are multiplying by 1,000,000, which is a power of 10, you can just move the decimal point, can't you? You move it once to the right per 0, you get $85,000 in interest on $1,000,000 a year, which you then add to your $1,000,000. Now you know why people who start with money keep getting more and more money.

8. 8.200015
   How It Works:
   You can do each set of parentheses however you want. The first (0.005 × 0.003) is probably easiest as decimals.

   0.000015

   $\left(\frac{1}{3} \times 24\right)$ is easy the way it's set up. It equals 8.

   $(20\% \times 1)$ you don't have to multiply, because multiplying by 1 is the same as doing nothing. Then you can leave it as a decimal.

   0.2

   Add them up along the decimal point and you have 8.200015.

9. She will give you 25 skateboards.

   How It Works:

   You can find out how many total skateboards she has to find out how many $\frac{1}{6}$ will be. 5 (suitcases) $\times 30$ (skateboards) $= 150$ skateboards.

   $$\frac{1}{6} \times 150 = 25$$

10. There has been a 99.98% decrease in minutes of math worry.

    How It Works:

    Percent decrease is the change over the original number. The change is $300,000 - 60 = 299,940$.
    The original number is 300,000.

    $$\frac{299,940}{300,000} \times 100 = 99.98$$

    There has been a 99.98% decrease in minutes of math worry. Good for you.

CHAPTER 3

RATIOS AND PROPORTIONS

Have you ever painted your room, or your apartment, or your face, and had a tough time choosing the color? One color might have too much orange, a different one might have too much blue. And for those of you who never thought about colors that way, what about when someone fixes you a peanut butter and jelly sandwich, or an ice cream sundae, or a martini? Sometimes there is too much fudge sauce, or too little jelly, or too much vermouth.

All of these situations revolve around **ratios**.

A ratio gives information about relative amounts within a whole. For instance, a perfect banana split—according to me--is 1 part banana, 3 parts ice cream, 2 parts toppings, and 1 part whipped cream.

This means the ratio of banana to ice cream to toppings to whipped cream is 1:3:2:1. These parts add up to 7.

Now, if you were looking at it as a fraction, you might say there are a total of 7 parts to the banana split. Banana is 1 part, so the fraction of the banana split that is banana is $\frac{1}{7}$.

That is a fraction in the form of part (the banana) to whole (banana split). Ratios, however, are expressed part to part.

The ratio of banana to ice cream in a banana split is 1:3. This means there is 1 part banana for every 3 parts of ice cream. There are other ways than the : to express ratios. Some people write them as words, "the ratio of banana to ice cream is 1 to 3." Or they write "1 part banana per every 3 parts ice cream." Do you remember per from percentages?

Ratios can also be written as division problems, the ratio of banana to ice cream is $1 \div 3$.

Other people write them so they look almost like fractions, the ratio of banana to ice cream is $\frac{1}{3}$. But remember, they are not fractions in the form of part to whole.

Fractions are expressions of **part to whole**. **Ratios** are expressions of **part to part**.

You should notice something. Here, as with all ratios you will encounter, order matters. Read a question carefully to determine which number you put first, with the : or the ÷ form, or on top with the / form. Whichever thing is listed first in the ratio, in this case bananas, goes on top, or first, and what is listed next goes on bottom, or next.

Mostly you will see ratios written with the colon, as in the ratio of banana to ice cream is 1:3.

Ratios don't have to be just integers either. You could have a ratio of $\frac{2}{3}$ a cup of flour to every $\frac{1}{4}$ cup of sugar, in a cake.

$$\frac{2}{3} \div \frac{1}{4}$$

To write this in its most reduced form, as you will be asked to write most ratios, just handle it the way you would handle a division problem with fractions.

$$\frac{2}{3} \times 4 = \frac{8}{3}$$

$\frac{8}{3}$ is the most reduced way to express the ratio of flour to sugar in the cake.

Take a look at a simpler ratio.

To make pink paint, you need 4 gallons of red, for every 1 gallon of white.

The ratio of red to white paint is 4:1 or $\frac{4}{1}$ or $4 \div 1$.

And it doesn't matter if you are measuring in gallons, pints, or tiny drops from an eye dropper. The relative amounts will always be the same. Just as with the banana split, the parts add up to the whole. The whole in this case, is 5. So, if you wanted to know what fractional part of the mixture was red, it would be the part that is red, or 4, out of the whole, or 5. The whole mixture is $\frac{4}{5}$ red, and $\frac{1}{5}$ white.

Since ratios are usually presented in their most reduced forms, just like fractions, the information above doesn't mean you only have 5 gallons of paint. It means whatever amount of this pink paint you have, the amount of red and white in that pink paint are related in the amount of 4:1.

Now say you have 100 gallons of paint, and you need to figure out what part is red and what part is white.

Since your ratio adds up to 5, figure out how many groups of this ratio will fit into your whole. Does this sound like the way division was explained? Well, it is.

$$100 \div 5 = 20$$

There are 20 groups of 5 in your final mixture. So you have 20 groups of your ratio. Which means your ratio has to be multiplied by 20 to find how many gallons of each type.

$$\frac{4}{1} \times \frac{20}{20} = \frac{80}{20}$$

You multiply both parts by 20, because you need to find out how many gallons of each type.

You have 80 gallons of red paint, and 20 gallons of white paint. Add them up and you have the 100 gallons of paint.

Another way to look at it is to remember that you already figured out what fractional part of the paint is red, $\frac{4}{5}$, and what fractional part was white, $\frac{1}{5}$. Since you know you have 100 gallons, to figure out what part is red you can just find $\frac{4}{5}$ of 100, and to figure out what part is white you can figure out $\frac{1}{5}$ of 100.

$$\frac{4}{5} \times 100 = 80$$

$$\frac{1}{5} \times 100 = 20$$

Again, you find you have 80 gallons of red and 20 gallons of white. And you add them up to check and you have 100 gallons of paint. The parts combine to form the whole.

The thing to remember about ratios is that they express part to part, and not part to whole.

Try it again with a more complicated ratio.

Each costume for a tacky Broadway show takes 3 parts sequins, 5 parts lace, and 7 parts spandex. They have 300 costumes, how many parts of lace did they use?

The ratio is 3:5:7 which adds up to a whole of 15.

This means each single costume is made up of 15 different parts.

So, in case you were interested, one costume is $\frac{5}{15}$, or $\frac{1}{3}$ lace. You can look at this in two ways.

One way is to realize that each costume is 15 parts, so 300 costumes would be 300×15 or 4,500 separate parts. Since you multiply the whole of 15 by 300, you multiply each part by 300.

$$5\ (\text{lace}) \times 300 = 1{,}500$$

You need 1,500 parts of lace.

Just to keep clear about it, you can see that if you multiply the other parts by 300, all your parts will add up to 4,500.

$$3\ (\text{sequins}) \times 300 = 900$$

$$7\ (\text{spandex}) \times 300 = 2{,}100$$

$$2{,}100 + 900 + 1{,}500 = 4{,}500$$

The other way to look at it is to say to yourself, "Hey, I will need 4,500 separate parts," by multiplying the whole, or 15, by 300. Since you know that fractionally, the lace accounts for $\frac{1}{3}$ of the parts, you could just find $\frac{1}{3}$ of 4,500.

$$\frac{1}{3} \times 4,500 = 1,500$$

Again, you get 1,500. It's all connected.

Some ratios won't add up to a whole, but they will still be consistently relative, as all ratios are. For instance, a flagpole that is 3 feet tall casts a shadow that is 2 feet long. The ratio of the height to the shadow cast is 3:2, but they do not combine to form a whole.

APPROXIMATE THIS:

What was the ratio of men to women in the U.S. Senate in October 2000?

Try to reduce the ratios that follow to their lowest forms. It may seem suspiciously like reducing fractions.

Quiz #1

1. 12:8
2. $\frac{4}{5}:\frac{1}{3}$
3. $\frac{\frac{2}{3}}{\frac{3}{5}}$
4. $\frac{8}{10}$
5. $\frac{1}{2}:4$
6. 18:9
7. $6:\frac{1}{2}$
8. 20:4

Proportions

One of the ways in which ratios are used is **proportion**.

A proportion is a way of comparing ratios with equal values. Proportion is usually used when figuring out how big or small things will be in comparable circumstances.

For instance, if the 3-foot-high flagpole casts a shadow of 2 feet, then a 6-foot-high sculpture will cast a shadow of 4 feet.

$$\frac{3}{2}=\frac{6}{4}$$

This is an example of **directly proportional**. As the thing casting the shadow gets bigger, the shadow also gets bigger. Directly proportional means both parts of the ratios being compared increase or decrease together.

Another way to write this expression is 3:2 :: 6:4. The four dots in the middle are the **proportion sign**. When a proportion is set up this way, the outer numbers, in this case 3 and 4, are called the **extremes**. The inner numbers, in this case 2 and 6, are called the **means**. One of the cool things about proportions is that since the ratios being compared are equal, the product of the extremes will equal the product of the means. Check it out and see.

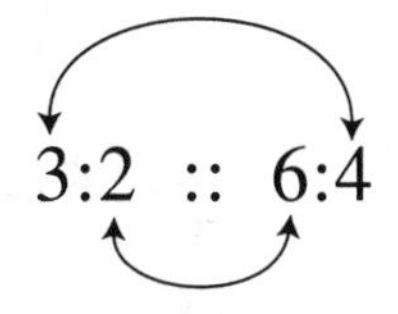

$$3\times4=12,\text{ and }\ 2\times6=12$$

How does this help you? By providing a way to figure out when a number is missing in a proportion. Remember, these are used to figure how an increase or decrease will affect things. Say you were told that for every 4 people who come to a birthday party, you need 3 bottles of soda. So people to soda is a ratio of 4:3. Then you found out you would have 16 people at this birthday party. How many bottles of soda will you need? Well of course the first thing you would do is approximate and write that down to compare with your answer later to see just how good at approximating you are getting. Then you set up your proportion.

$$4:3 :: 16:?$$

You are missing one of the extremes, but don't worry. Since you cannot multiply the extremes, multiply the means.

$$3 \times 16 = 48$$

This means the extremes will have to multiply to be 48 as well.

$$4 \times ? = 48$$

What you can do to find the missing piece is to divide 48 by 4. Remember, division is the inverse of multiplication.

$$48 \div 4 = 12$$

To test it out, just put 12 into your proportion.

$$4:3 :: 16:12$$

Do the extremes multiply and yield the same product as the means? You're in business. And you'll buy 12 bottles of soda for the party.

This works when you have proportions set up other ways as well.

$$\begin{matrix} \text{people} \longrightarrow \\ \text{soda} \longrightarrow \end{matrix} \frac{4}{3} = \frac{16}{12} \begin{matrix} \longleftarrow \text{people} \\ \longleftarrow \text{soda} \end{matrix}$$

In this case, you have the same proportion set with the ratios in fraction bar form, across an equal sign. It is still a proportion because you have set the two ratios as equal. Here, the extremes are **diagonally** opposite to one another. Diagonal means across on a tilted axis. But, you still multiply them and it works out just the same. This is called **cross-multiplying**. However, just because it may remind you in some fond way of cancelling, do not think, that it is cancelling.

YOU CAN NEVER CANCEL ACROSS AN EQUAL SIGN.

$$3 \times 16 = 48, \text{ and } 4 \times ? = 48$$

When you are writing proportions this way, make sure you are putting similar values in the same place. In this case, people go on top, and soda goes on bottom.

$$\frac{4}{3} = \frac{16}{12}$$

As long as you are careful of that, proportions will be great. You could also write it with people on one side, and soda on the other, as long as you remember to keep it small to big.

$$\text{people}\left\{\begin{matrix}\text{small} \longrightarrow \\ \text{big} \longrightarrow\end{matrix}\right. \frac{4}{16} = \frac{3}{?} \left.\begin{matrix}\longleftarrow \text{small} \\ \longleftarrow \text{big}\end{matrix}\right\}\text{soda}$$

Try these proportion questions:

Quiz #2

1. In the proportion ?:3 :: 6:2, what is the missing piece?
2. In the proportion 4:1 :: ?:2, what is the missing piece?
3. In the proportion $\frac{5}{6} = \frac{25}{?}$, what is the missing piece?
4. In the proportion $\frac{3}{?} = \frac{9}{3}$, what is the missing piece?
5. In the proportion $1:\frac{1}{2}$ :: 10:?, what is the missing piece?
6. In the proportion $\frac{?}{4} = \frac{75}{100}$, what is the missing piece?
7. In the proportion 2:? :: 5:10, what is the missing piece?
8. In the proportion $\frac{4}{5} = \frac{?}{35}$, what is the missing piece?

So, did you find yourself thinking they were getting easier and easier as you went along? Do you see connections with reducing fractions and common denominators? The more you practice math, and write down all the pieces of it, and turn it around in your head, the more math smart you become.

INVERSE PROPORTIONS

There is another type of proportion that can sometimes be a bit daunting. This is when things are **inversely proportional**. Inversely proportional means that when one part of your ratio increases, the other part decreases. For instance, the more time I spend worrying, the less work I do. If I worry for 1 hour, I get 6 hours of work done. But if I worry for 2 hours, I only get 3 hours of work done. It makes sense, right? The more time I spend worrying, the less I work.

Since these are inversely proportional, you will have to invert the ratios. The easiest way to do this is to put worry on one side, and work on the other.

$$\frac{1 \text{ worry}}{2 \text{ worry}} = \frac{3 \text{ work}}{6 \text{ work}}$$

What you are probably wondering is, "Why is the second one in the upside down position? When you worry 1 hour you work 6 hours, not 3 hours."

When you set up an inverse proportion, you need to invert one of your ratios, and it is easy to remember to invert the second ratio. When you cross multiply in a regular proportion, the products are equal. It works the same way for an inverse proportion.

Look at an example where you need to find a missing piece.

There are 4 women building a house. It takes these 4 women, who, coincidentally all work at the exact same rate, 20 days to build a house. Now, if all of a sudden 4 more women who worked at the exact same rate came to help, it would end up taking less time, right? More workers, less time—an excellent example of inversely proportional. So set up your proportion.

$$\frac{4 \text{ women}}{8 \text{ women}} = \frac{? \text{ days}}{20 \text{ days}}$$

Remember, you invert the second part. Now you can cross-multiply.

$$4 \times 20 = 80$$
$$8 \times ? = 80$$

You divide 80 by 8 and get 10. It will take 10 days if there are 8 women working instead of only 4. Pretty reasonable stuff, right?

APPROXIMATE THIS:
What was the ratio of men to women in the U.S. Senate in January 1993?

GLOSSARY

ratio: A relationship between two quantities.
proportion: A relationship between two ratios.
directly proportional: A proportion in which the quantities in the ratios increase or decrease at the same time.
inversely proportional: A proportion in which, within a ratio, one quantity increasing causes the other quantity to decrease.
extremes: The outer terms in a proportion.
means: The inner terms in a proportion.

What you just read is basically all there is to ratios and proportions, and now you can combine the pieces of knowledge and use them on the following problems. Take your time, relax, and as always, approximate before you work the problems out. Approximating is one of the few things that make life more sensible instead of less.

Quiz #3

1. How many dogs are there in a group of 200 pets if there are only dogs and cats in the group, and the ratio of dogs to cats is 3:1?
2. A good party requires a mixture of funny people and dancing in a ratio of 1:5. If there are 20 funny people who will come to Betsy's party, how much dancing does she need to make it a good party?

3. A fruit bowl's mixture is fractionally $\frac{1}{3}$ apples and $\frac{2}{3}$ oranges. What is the ratio of oranges to apples in the fruit bowl?
4. There is a museum of art works that has only sculptures, paintings, and music. The ratio of sculptures to paintings to musical compositions is 3:4:5. If there are exactly 36 paintings in the museum, how many art works are there in the museum total?
5. If there are 270 students in Mr. Peck's class, and the ratio of boys to girls is 2:1, how many boys are there in his class?
6. If it takes 10 people to paint a house in 7 hours, how long will it take 5 people to paint the same house if they are all painting at the same rate?
7. If there are 50 hats in Jane's closet, and 10 of them are cowboy hats, and the other 40 are bowler hats, what is the ratio of cowboy hats to bowler hats in Jane's closet?
8. There is a bookstore that has only novels and dictionaries, and the ratio of novels to dictionaries is 5:7. What fractional part of the books in the bookstore are dictionaries?
9. If it takes 20 people to carry a statue that weighs 500 pounds, how many people will it take to carry a statue that weighs 1,500 pounds?
10. If it takes Janice 4 hours to travel the 10 blocks from her house to her high school, how long will it take her, traveling the same rate per block, to travel the 35 blocks from her house to the gym?

Way to go! Let's press on.

ANSWER KEY

Quiz #1

1. 3:2
2. 12:5
3. $\frac{10}{9}$
4. $\frac{4}{5}$
5. 1:8
6. 2:1
7. 12:1
8. 5:1

Quiz #2

1. 9
2. 8
3. 30
4. 1
5. 5
6. 3
7. 4
8. 28

Quiz #3

1. 150 dogs
   How It Works:
   The whole of your ratio here is 4, and you need to multiply it by 50 to get 200 pets. $200 \div 4 = 50$
   Since dogs are 3 parts of the whole, you multiply the 3 by the 50 as well, and get 150 dogs.

2. 100 parts dancing
   How It Works:
   The ratio of people to dancing is 1:5 and they want the proportional number of dancing for 20 people.

   $$1:5 :: 20:?$$
   $$5 \times 20 = 100$$
   $$1 \times ? = 100$$

3. 2:1
   How It Works:
   The whole is 3, and the parts are 1 part apples to 2 parts oranges, so remember they are asking for the ratio of oranges to apples. Oranges come first. 2:1

4. 108 art works
   How It Works:
   There is a ratio of 3:4:5 of sculptures to painting to music. This gives your ratio a total of 12 pieces of art work. You find out you have 36 paintings. What did you multiply to get 36? Divide and find out.

   $$36 \div 4 = 9$$

   So you also need to multiply your total by 9 to find how many pieces of artwork there are total.

   $$12 \times 9 = 108$$

5. 180 boys
   How It Works:
   The ratio is 2:1, and that means the total of your ratio is 3. Use the fraction for a change. What fractional part is boys? $\frac{2}{3}$.

   So $\frac{2}{3}$ of 270 are boys.

   $$\frac{2}{3} \times 270 = 180$$

6. 14 hours
   How It Works:
   Excellent! An inverse proportion. Fewer people will take more time. Put people on one side, and invert the time.

   $$\frac{10}{5} = \frac{?}{7}$$
   $$10 \times 7 = 70 \quad 5 \times ? = 70$$
   $$70 \div 5 = 14$$

7. 1:4
   How It Works:
   10:40 reduces to 1:4

8. $\frac{7}{12}$

   The total of your ratio is $5 + 7 = 12$. Of this total, 7 parts are dictionaries, so $\frac{7}{12}$ of the books are dictionaries.

9. 60 people
   How It Works:
   Set up your ratios as proportions, and find the missing piece.

   $$20:500 :: ?:1{,}500$$
   $$20 \times 1{,}500 = 30{,}000$$
   $$500 \times ? = 30{,}000$$
   $$30{,}000 \div 500 = 60$$

10. 14 hours
    How It Works:
    Another proportion, so 4:10 :: ?:35

    $$4 \times 35 = 140$$
    $$10 \times ? = 140$$
    $$140 \div 10 = 14$$

CHAPTER 4

EXPONENTS AND ROOTS

You remember exponents from the first chapter. Exponents are a shorthand way of expressing a number multiplied by itself. For instance, 4^3 is 4 times itself three times.

$$4 \times 4 \times 4 = 64$$

The way to express 4^3 aloud is to say either "four to the third power" or "four cubed."

When a fraction is being raised to an exponential power, both the numerator and the denominator get raised to that power. For instance, $\left(\frac{3}{4}\right)^2 = \frac{3 \times 3}{4 \times 4} = \frac{9}{16}$.

Other exponent facts you might want to remember are on page 47. Now for some new and fancy exponent information.

Multiplying Exponents

What if you wanted to multiply 4^3 by 4^2? How would you express that using exponents? Take a look at what each of those exponents means.

$4^3 = 4 \times 4 \times 4$, and $4^2 = 4 \times 4$, so $4^3 \times 4^2 = 4 \times 4 \times 4 \times 4 \times 4$. There are 5 fours, so you have 4^5. Since you see how it works, and now it makes perfect sense, from here on in you can just remember, **when multiplying exponents with the same *base*, just add the exponents**. And what the heck is the base, you say? It is the number that is being raised to a power. In this particular case, the base is 4.

Dividing Exponents

What about when you are trying—you are getting very ambitious here—to divide exponents? For instance, $\frac{3^5}{3^2}$.

Again, just as you did with multiplying exponents, here you should arrange them as multiplication and see what happens.

$$\frac{3 \times 3 \times 3 \times 3 \times 3}{3 \times 3}$$

Two of the threes cancel out, correct? Both $\frac{3}{3}$ and $\frac{3}{3}$ get counted as 1, because any number over itself is equal to 1 and they both disappear. You are left with $3 \times 3 \times 3$. Or, 3^3. So the rule here is that **when you divide exponents with the same base, subtract the exponents.**

"But," you ask, "what would have happened if the smaller exponent had been on top, and the larger exponent had been on the bottom?" Well, generally you would leave the larger where it is. $\frac{2^4}{2^7} = \frac{2 \times 2 \times 2 \times 2}{2 \times 2 \times 2 \times 2 \times 2 \times 2 \times 2}$

Cancel where the $\frac{2}{2}$ appears, and you are left with $\frac{1}{2 \times 2 \times 2}$ or $\frac{1}{2^3}$. So you subtract smaller from larger.

But can you subtract and get 2^{-3}?

NEGATIVE EXPONENTS

What happens when you raise 2 to the negative third power (2^{-3})? It is the same as if you had left it $\frac{1}{2^3}$. It becomes $\frac{1}{2 \times 2 \times 2}$, or $\frac{1}{8}$.

Any number raised to a negative power becomes 1 over that number to its exponent's positive power. It all connects.

APPROXIMATE THIS:

If on the first day of the year, someone gave you \$2, and from then on they said they would double the amount of money every day for the whole month of January, how much money would you have on January 31?

ADDING AND SUBTRACTING EXPONENTS

One of the sad things in this world is that there is no real way to add or subtract exponents. Think about it. If you have $2^3 + 2^4$ it is equal to $(2 \times 2 \times 2) + (2 \times 2 \times 2 \times 2) = 8 + 16 = 24$. Unfortunately, you can see that there is no special exponential shortcut, since 24 is not any power of 2, $2^4 = 16$, and $2^5 = 32$.

If you have $2^3 + 2^3$ you can say it is $2(2^3)$, but that is regular old multiplication. Two groups of 2^3 is the same as two times 2^3. Which is 16, by the way. But other than that, there is no way to add exponents with the same base.

RAISING EXPONENTS TO OTHER POWERS

How would you react when faced with something like 4^{2^3}? To see how it works, try expanding it out.

$$4 \times 4 \ \times \ 4 \times 4 \ \times \ 4 \times 4 \ = 4^6$$

When you have an exponent being raised to another exponent, you can multiply the exponents.

What you need to remember with all of this is that the easiest way to see and understand the way exponents work is to write them out. So if you are stuck, and you are thinking to yourself, "What's that darn rule? Is it multiply? Or maybe it's add?" just write out the exponents and count up the pieces yourself. Understanding the ways in which math functions is infinitely more helpful than memorizing a bunch of rules without knowing what they mean.

SCIENTIFIC NOTATION

Numbers can get really huge. Really, really huge. Remember a few pages back when π took up 22 decimal spaces? While that may be fun to do when you are expressing just how long numbers can be, it's a tremendous waste of time and paper to do that often. A way to get around it is something called **scientific notation**. Scientific notation uses exponential powers of ten as a way of expressing numbers.

$$56.7 \times 10^2$$

Here is an example of scientific notation. Again, to figure out what is going on, expand it. What is 10^2? 100. So you have 56.7×100. You know that to multiply by 100 you move the decimal point two spaces to the right, so $56.7 \times 100 = 5,670$.

That is how scientific notation works. Now that you understand it, here is a simpler way to translate the pieces.

Check the exponent of the 10, then move the decimal point that many places to the *right* if the exponent is *positive.*

$$56.7 \times 10^2 = 56.70$$

The exponent here is 2, so you move the decimal point two spaces to the right to get 5,670.

If you get a scientific notation expression in which the exponent is negative, remember what a negative exponent means. $10^{-2} = \frac{1}{100}$. Find the exponent of the ten, and move the decimal point that many places to the *left*.

$$5.67 \times 10^{-2} = 0.0567$$

Roots

You know that if you square 3, you get 9.

$$3^2 = 3 \times 3 = 9$$

And you know that if you square -3, you also get 9.

$$-3^2 = -3 \times -3 = 9$$

But what about the famed **square roots**? A square root of a particular number is a number that, when squared, will equal that particular number. For example, the square root of 9 is 3.

Why can't it be both 3 and -3? Because the definition of square root is that it always a positive number.

The sign for square root looks like this: $\sqrt{\ }$.

$\sqrt{25}=$ the square root of 25. You need to ask yourself, "What times itself equals 25?" And the answer is 5.

9 and 25 are what is known as **perfect squares**. This means they are the squares of integers. There are other numbers you will come across that are not perfect squares. For instance, $\sqrt{8}$ is not equal to an integer. This means 8 is not a perfect square. You can approximate, of course, and you should, so you feel comfortable with it. Is $\sqrt{8}$ bigger than 2? Well yes, if it were 2, 2^2 is equal to 4, and 8 is bigger than 4, so $\sqrt{8}$ is bigger than 2. Is it bigger than 3? Well, 3^2 is equal to 9, and 9 is bigger than 8, so $\sqrt{8}$ is smaller than 3. In fact, $\sqrt{8}$ is between 2 and 3.

Another way to handle square roots of numbers that are not perfect squares is to factor them out. This means, find the factors of the number you are taking the square root of, and see if you can take a square root of these factors.

$$\sqrt{8}=\sqrt{4\times 2}$$

Well, 4 is a perfect square, and the square root of 4 is 2. So, you put the 2 outside the square root sign, and leave what remains that cannot be factored to a perfect square, inside. It looks like this,

$$\sqrt{8}=\sqrt{4\times 2}=2\sqrt{2}$$

This also shows you that you can multiply square roots by each other. $\sqrt{4}\times\sqrt{2}=\sqrt{8}$.

OTHER ROOTS

Square roots are not the only type of roots there are; there are as many other kinds of roots as there are numbers. Yikes.

To figure out what kind of root a number is, take a look at the number floating high on the left of the **radical sign**. Radical is another way to say square root sign.

$\sqrt[4]{16}$ = the fourth root of 16. This means some number to the fourth power is equal to 16. Here, the number is 2, because $2 \times 2 \times 2 \times 2 = 2^4 = 16$.

Don't worry though, the ones you will come across most frequently are square roots and **cube roots**. A cube root looks like this: $\sqrt[3]{\ }$.

The cube root of a particular number is a number that when cubed, equals the particular number. Whew. For instance, the cube root of 8?

$$\sqrt[3]{8} = 2, \text{ because } 2 \times 2 \times 2 = 2^3 = 8.$$

Just like with square roots, cube roots are only positive numbers. And just like square roots, and in fact, everything in the world, you should always try to approximate cube roots so you feel comfortable with them as numbers of a certain type.

$$\sqrt[3]{9}$$

Is it bigger than 2? Well, $2^3 = 8$, and 9 is bigger than 8, so yes, $\sqrt[3]{9}$ is bigger than 2.

Is it bigger than 3? Well, $3^3 = 27$, and 9 is not bigger than 27, so no, $\sqrt[3]{9}$ is a lot smaller than 3, it is some number between 2 and 3, that is closer to 2.

Another way to express roots is as fractional exponents. $27^{\frac{1}{3}}$ is just another way of expressing $\sqrt[3]{27}$, or the cube root of 27, which equals 3. And $36^{\frac{1}{2}}$ is another way of expressing $\sqrt{36}$, or the square root of 36, which equals 6.

GLOSSARY

base: A number being raised to an exponential power.

scientific notation: A method of expressing very large or very small numbers by multiplying by powers of 10.

square root: A factor of a number that, when squared, equals the number.

perfect square: A number that is the square of an integer.

radical sign: A sign indicating a root of some kind, it looks like this: $\sqrt{\ }$.

cube roots: A factor of a number that, when cubed, equals the number in question.

Now look at a bunch of problems using exponents and powers.

Quiz #1

1. 5.6×10^6
2. $4^{-6} \times 4^8$
3. $16^{-2} =$
4. $\frac{8^4}{8^2} =$
5. $\sqrt{49} =$
6. $4.5 \times 10^{-4} =$
7. $4^4 =$
8. $5^{-2} =$
9. Factor out $\sqrt{50}$
10. $\sqrt[3]{64} =$

ANSWER KEY

Quiz #1

1. 5,600,000
   How It Works:
   The exponent on the 10 is 6, so you move the decimal point six places to the right.

2. 4^2 or 16
   How It Works:
   When multiplying exponents with the same base, like these, add the exponents.
   $-6+8=2$, so you have 4^2 or 16.

3. $\frac{1}{256}$
   How It Works:
   When you have a number raised to a negative exponent, you put that number to the same exponent, but positive, under 1. So here you get $\frac{1}{16^2}$ or $\frac{1}{256}$.

4. 64, or 8^2
   How It Works:
   When you divide exponents with the same base, as you do in this case, you subtract the exponents. It's like canceling.

$$\frac{8\times8\times8\times8}{8\times8}$$

   So you have 8×8 which equals 64.

5. 7
   How It Works:
   You are being asked, "What, times itself, would equal 49?" Try integers. Is it 5? $5\times5=25$, so no. It's bigger, so try 7. $7\times7=49$, so $\sqrt{49}=7$.

6. 0.00045
   How It Works:
   This number is being expressed in scientific notation. Since the exponent on the 10 is negative, you need to move the decimal point to the left. Since the exponent is -4, you need to move the decimal point four places to the left.

   .0004.5

7. 256
How It Works:

$$4^4 = 4 \times 4 \times 4 \times 4$$

Multiply it out, you get 256.

8. $\frac{1}{25}$
How It Works:

5^{-2} means you need to invert the base and then do the exponent.

$$\frac{1}{5^2} = \frac{1}{5 \times 5} = \frac{1}{25}$$

9. $5\sqrt{2}$
How It Works:
To factor out a square root, you need to see if there are any perfect squares in the factors of the number.

$$\sqrt{50} = \sqrt{25} \times 2$$

Since 25 is a perfect square, you can factor out its square root. What, times itself, equals 25? 5, so you put the 5 outside the radical sign, and leave the 2 inside.

$5\sqrt{2}$. Now, can you factor the 2? No, the 2 is prime, and it has no perfect square, so you are done.

10. 4
How It Works:

$\sqrt[3]{64}$ is asking you what number, cubed, will equal 64? Fish around, approximate. Is it 2? $2 \times 2 \times 2 = 8$, too small. How about 3? $3 \times 3 \times 3 = 27$, still too small, how about 4? $4 \times 4 \times 4 = 64$. Bingo.

CHAPTER 5

ALGEBRA

Math is used to relate numbers and quantities and their relationships. So far in this book, you have been using math to figure out amounts. For instance, how many pears does John have, how many hats are in Emily's closet, things like that.

There is another way to use math, for a slightly different purpose—to describe mathematical relationships in a more abstract way. It is called, get ready now, **algebra**. Algebra is math that uses letters to represent numbers.

These letters are referred to as **variables**. It makes sense, because it means that the value for a letter varies; it isn't fixed. When variables are combined and set equal to other variables or numbers, it is called an **equation**.

A **formula** is a specific type of equation that is used to represent a particular relationship between things.

Say I wanted to tell you that for every gift you give me, I will thank you 3 times. Use g for gift, and t for thank you. We'll call it the politeness formula. You can use algebra to write a formula that expresses this relationship in math form.

$$3g = t$$

This means 3 times the number of gifts equals the number of thank you's. The 3 goes with the gifts because this is set up as an equation, which means that both sides are equal. To make them equal in this case, since according to the formula there are more thank you's than gifts, we had to multiply the gifts by 3 to equal the thank you's. For instance, 1 gift means 3 thank you's. To set them equal to each other, we need to multiply the gift by 3.

And how does $3g$ indicate three times the number of gifts? Do you remember that numbers in parentheses next to each other means multiply? Like $(3)(-4) = -12$.

In algebra you can put the variables right next to each other, or numbers right next to a variable, like $3g$, and it means multiply. And by the way, in algebra, whenever a number is being multiplied by a variable, that number is called the **coefficient**. In this case, g has a coefficient of 3. When a variable appears to stand alone, this means it has a coefficient of 1.

coefficient

$$3g = t$$

Another example of a formula is the distance formula. The rate, or speed, at which you drive, multiplied by the time driven, in hours, will give you the distance driven. The way to express this relationship between rate, time, and distance is

$$\mathrm{R} \times \mathrm{T} = \mathrm{D}$$

R stands for rate, T stands for time, and D stands for distance. You can use algebra to express your own made-up formulas (like the gift-to-thank you relationship), or to keep in mind formulas that are always true and you might need to remember sometime (like $\mathrm{R} \times \mathrm{T} = \mathrm{D}$).

Writing Your Own Formulas

To make an algebraic formula to represent a particular relationship, supply a variable for each part of the relationship, and then translate that relationship into math language.

Translating Thoughts

Remember when fractions were discussed, and you read that "$\frac{1}{2}$ of 4" meant $\frac{1}{2}$ times 4? Well these sorts of translations are what you work with in algebra.

"Of" translates to multiplied by.

"More" or "greater than" means add.

"Less" or "less than" means subtract.

"Is" means equals.

Most translations are pretty logical. If you have trouble figuring out what something means, try substituting in numbers.

For instance, if a question says, *"Q is three less than V" say to yourself, if Q were any number at all, say 4, what would V be? 7.*

So Q is smaller than V, and it is exactly three numbers smaller. $Q = V - 3$.

Try another one. John is 5 years older than Mary.

Well, John is older by 5 years, so you want to first say John is older, and Mary is younger. This means to even out the equation, you will need to add 5 years to the side with Mary.

$$J = M + 5.$$

Test it out, if Mary is 30, this makes John 35. Is John 5 years older than Mary? Yes.

Try transforming the following relationships into algebraic formulas. You can use whichever variables make the most sense to you, just make sure you keep them carefully labeled on your paper so it's not a nightmare to check them against the answer key.

Quiz #1

1. The number of girls in a room is equal to half the number of boys.
2. Sales tax is equal to 8% of the price.
3. A pair of shoes costs four times as much as a hat.
4. Stanley is 4 years younger than Margaret.
5. One computer is equal to 3 typewriters and 4 calculators.
6. George's weight is one third of Ben's weight.
7. David is 3 years older than Hanna.
8. Three books are received for every library visit.

Solving for X

Formulas are used to solve problems. You can put in numbers for the variables and find out information.

For instance, in the politeness formula, $3g = t$, you can now figure out how many thank you's are necessary for a particular number of gifts. It is called **substituting in**, or **evaluating an equation**.

If I gave you 5 gifts, how many thank you's would you need to produce?

Substitute in the 5 for the g, and calculate.

$$3 \times 5 = t$$

$$15 = t$$

You need 15 thank yous.

You can also substitute in for thank you's. Say you knew, somehow, that someone was acting according to the politeness formula. And you saw them at a party and they thanked some person 27 times. You want to figure out how much loot they scored so you substitute 27 into your formula for the thank you variable.

$$3g = 27$$

Now you need to **isolate the variable**. This means you need to get the g by itself so you know its value. The way you do this is to move the numbers to one side of the equation, and the variable to the other. You see how equations have two sides? The left side of this equation is $3g$, and the right side of this equation is 27. The two sides are separated by the equals sign. You want to end up with the numbers on one side of the equation, and the variables on the other.

The thing about equations is, you can do anything you want to them, *as long as you do it to all parts and both sides of the equation.*

WHY THIS WORKS

Any equation is telling you that one side is equal to the other side. In this case, $3g$ is equal to 27. So if you added 15 to both sides, you wouldn't be changing it at all, since you are doing equal things to both sides. You would have $3g + 15 = 42$. The two sides would still be equal, and g would still have the same value. Read on to see how that's true.

Back to isolating the variable. In the case of $3g = 27$, the coefficient 3 needs to be moved in order to obtain the actual value of g.

The way to move something from one side of the equation to the other is to perform the inverse operation to both sides of the equation. In this case, the 3 is being multiplied by g, so this means you need to divide by 3. Again, the most important thing is that whatever you do to the equation is okay, as long as you do it to both sides of the equation. So divide away.

$$\frac{3g}{3} = \frac{27}{3}$$

Use the fraction bar to show division.

$$g = 9$$

The guy just got 9 gifts.

So how does this dividing with variables work?

When you perform the inverse operation, in this case divide, you have $\frac{3g}{3}$. Well, you remember $\frac{3}{3}$, it's 1, so you end up with $g \times 1$, or g, equal to $\frac{27}{3}$, or 9. When you divide and multiply by the same number, it is the same as multiplying by the number over itself, in this case, $\frac{3}{3}$. And you know that multiplying by any number over itself is multiplying by 1.

You can substitute back to check your work.

$$3 \times 9 = 27$$

Now you can check and see if the earlier explanation works.

$$3 \times 9 + 15 = 42$$

Pretty smooth, right?

Try another example.

Five less than a certain number is equal to 20.

The way to write this as an equation is $x - 5 = 20$.

When you are asked for "a certain number," that number is a variable, or the unknown. You can write n or a or x, or any variable you like. Once you do that, you are ready for the next step. Now you have to solve your equation. You need to isolate the variable.

$$x - 5 = 20$$

To get x by itself, which is really what isolating means, you want to move that pesky -5. How to do it?

Remember, the way to move a number is to perform the opposite operation. The inverse operation of subtraction is addition. Since this is asking you to subtract 5, you need to add 5. If you subtract 5 and then add 5, it is the same as doing nothing at all, since they cancel each other out to 0. Look at an example of adding a number and then subtracting it.

$$7+3-3=7$$

So adding the 5 leaves x by itself. But remember, no matter what operation you do, you must do it to both sides of the equation to keep all things equal.

$$x-5+5=20+5$$

$$x=25$$

If you want to be sure that this is truly the value of x, you can substitute it back into your formula.

$$25-5=20$$

Looks pretty good.

Sometimes you will use substituting to check your work, like here where you found your own value for x. Other times you will have a formula (like you did with the politeness formula) and you will be given a value for one of your variables. Try that type of substitution again.

Alice is 3 years older than Bob. If Bob is 14, how old is Alice?

So you write, $A=B+3$, to represent that Alice is 3 years older than Bob.

Then the question asks, if Bob is 14, how old is Alice?

And you substitute the 14 for Bob.

$$A=14+3$$

$$A=17$$

Alice is 17 years old.

Sometimes the equations will get a little bit more complicated, and have variables appearing on both sides, but don't let it throw you. Just collect the variables on one side, and the numbers on the other.

Five more than a certain number is equal to twice that number.

How do you translate that? "Five more" means add 5, and "twice" means times 2, and as you remember, "a certain number" means a variable and "is" means equals.

$$x + 5 = 2x$$

You want all the variables together. One way to think of it is as a balance. On which side are there already more variables? On the right side there are two *x*s, and on the left side there is one. So you might as well continue the trend, and put all the variables on the right side. It makes no difference really, just do whichever is easier for you on any particular problem. How do you get all the variables on one side?

Well on the left side there is $x + 5$. Since in the equation the x is being added, to move it to the other side, you need to subtract it. And, as always, you must do whatever you are doing to both sides of the equation.

$$x + 5 - x = 2x - x$$

$$5 = x$$

The $2x - x = x$ is because 2 times x really means you have two x terms. If you have two variables and then you subtract one away, you are left with only one. To check your answer, try substituting your solution, 5, for x.

$$5 + 5 = 2 \times 5$$

$$10 = 10$$

In the following problems, write out the equations and then solve for x.

Quiz #2

1. Four more than a certain number is equal to 7.
2. Twice a certain number is equal to that number minus 3.
3. A certain number plus 6 equals 15.
4. Three times a certain number equals that number plus 14.
5. A certain number plus 3 is equal to 40 minus that same number.
6. A certain number divided by 8 is equal to 6.
7. Forty minus a certain number, is equal to 3 times that number.
8. A certain number plus 9 is equal to that number times 2, minus 3.

What you have just been doing is solving for the variable in algebraic expressions. That's one of the most concrete aspects of algebra. It helps you find missing pieces in a huge number of problems. Most of your unknown pieces can now be found by moving around your equations. This is called **manipulating the equation**.

Solving in Terms of Variables

Other parts of algebra call for more abstract kinds of things.

For instance, you may be asked to solve for a variable in an algebraic equation in terms of another variable. This means the answer will not be a number, but rather a definition of the variable in question, in terms of other variables. Here's what it will look like.

$$3x + 7 = 2y$$

Give x in terms of y.

This means you are again being asked to isolate x. Since you want x on one side of the equation, you need to move the 7 by subtracting it from both sides of the equation.

$$3x + 7 - 7 = 2y - 7$$

$$3x = 2y - 7$$

Now you need to move the 3, so x stands alone. Since x is being multiplied by 3, divide both sides by 3.

$$\frac{3x}{3} = \frac{2y - 7}{3}$$

$$x = \frac{2y - 7}{3}$$

So you now have a way of expressing x in terms of y. Try expressing these equations in terms of variables:

Quiz #3

1. $3a - b = 4$ Express a in terms of b.
2. $x + 2y = 3x$ Express y in terms of x.
3. $w = z - 4$ Express z in terms of w.
4. $3e = f$ Express e in terms of f.
5. $a + 7 = b - 6$ Express b in terms of a.
6. $x + y = 5$ Express x in terms of y.
7. $a = 3b + 2$ Express b in terms of a.
8. $2c - 2d = 8$ Express c in terms of d.

DEFINITIONS

In these sorts of algebraic expressions, you need to learn a few more definitions.

You already know that variables are used to represent numbers. Remember in ratios and proportions when you had to find the missing piece? Variables can represent the missing pieces. Sometimes people will also call a variable the **unknown**. You can sort of imagine all these mathematicians staying up too late one night and maybe having a few drinks and saying, "Wow. The unknown." But in algebra, the unknown is knowable!

Pretty much, anyway, in the form of a numerical value or in terms of another variable.

$$6 + x = 10$$

In the equation above, x is called a variable or an unknown, and $6 + x$ is called an **expression**.

An expression is a bunch of numbers and variables being added or subtracted. You have already seen expressions in the formulas and equations you have been working with. The numbers and variables can be combined by multiplication or division, or just left alone. Those numbers and variables that combine to form an expression are called **terms**.

In $6 + x = 10$, 6 is a term, and x is a term. Take a look at another expression.

$$\frac{x}{4} - xy + 3y$$

The terms in this expression are $\frac{x}{4}$, xy, and $3y$. Terms are numbers, variables, or numbers and variables that are combined by multiplication or division.

You would not call $\frac{x}{4} + 3y$ a term, because a term includes only variables and their coefficients—the numbers that are being multiplied or divided. Remember, dividing a variable by 4 is the same thing as multiplying it by $\frac{1}{4}$. With addition, the components are recognized as separate terms.

Not all these expressions will be in the form of equations. Sometimes they will not be set equal to anything. They will just be standing there and you will need to make them clearer, or simplify them.

For the next few questions, count up how many terms there are, and what the coefficient is in each term, if it has one. Remember coefficients from page 160. Try one here for practice.

$$5x - 2y^3 + 6$$

There are 3 terms, the first term is $5x$, the second term is $2y^3$, and the third term is 6.

Yes, 6 is a term, terms can be numbers or variables or combinations of numbers and variables that are multiplied or divided.

For the first term, 5 is the coefficient, because the x is multiplied by 5. For the second term the coefficient is 2, because y^3 is multiplied by 2. And the third term, 6, has no coefficient, since there is no variable, and a coefficient is the number by which a variable is multiplied. Now, you try it.

Quiz #4

1. $7x + 6xy$
2. $-4x^2 + 3y + 5$
3. $7a - 3ab - b^2$
4. $3c - b^3 - 7$
5. $4x + 4y$
6. $3a + 2b$
7. $4a^2 + 2ab + b^2$
8. $2x + 5y$

WAYS OF SIMPLIFYING

Simplifying means putting an expression into simpler terms. It's a little like reducing a fraction. The expression will have the same value, it will just be a little easier to handle when it is simplified.

SIMPLIFYING BY COMBINING TERMS

One way to simplify is to combine terms. When you have an expression containing **like terms**, you can combine them. Like terms are elements of an expression that use identical variables. You have already combined like terms when you solved equations that had variables on both sides. It is a good idea to combine whenever you see like terms.

$$7x + 4x + y \text{ will combine to } 11x + y$$

In any expression, x represents one thing the whole way through, so you can combine all the information about x. But you cannot combine the information about two unlike terms.

$$7x + 3y \text{ stays } 7x + 3y$$

Think of x as representing hats and y as representing sweaters. The items 7 hats and 3 sweaters do not combine to form 10 hats and 10 sweaters. So combine only like terms. Or think of x and y as representing actual numbers. Try substituting.

For $7x + 3y$, if $x = 2$, and $y = 3$, then $7 \times 2 + 3 \times 3 = 23$.

But $10xy$ would equal $10 \times 2 \times 3 = 60$, if you were to combine like terms.

Never try to combine unlike terms. And just to make things a little more interesting, like variables with unlike exponents are NOT considered like terms. You cannot combine $3x^2$ and x^3. This is because, $3 \times x \times x + x \times x \times x$ does not equal $3x^5$, does it? You cannot add exponents. If you have a question about this, go back to page 47 in chapter 4.

Combining terms can make long terrifying expressions seem much more manageable. Take a look at this one.

$$5x + 7y - z - 2x + 4y - 2z + 10x$$

Just approach it one step at a time. Take the x terms first, since they come first. There is $5x$, there is $2x$, and there is $10x$. These are all like terms.

The $5x$ and the $10x$ are being added in the long run, but the $2x$ is being subtracted. Well fine, just add the $5x$, subtract the $2x$, and add the $10x$.

$$5x - 2x = 3x \text{ and } 3x + 10x = 13x$$

You have $13x + 7y - z + 4y - 2z$.

Now take the y terms. You have $7y$ and $4y$. These are like terms.

$$7y + 4y = 11y$$

You've gotten it down to $13x + 11y - z - 2z$.

Now the z terms. One thing you should notice is that most algebraic expressions like to have the variables in alphabetical order. It's sort of sweet, really. It's the meeting of English and math. You have a z being subtracted, and then $2z$ being subtracted. Well, luckily, you know how to combine two negatives.

$$-z - 2z = -3z$$

Now you can string all your combined terms together into an expression of three simple terms.

$$13x + 11y - 3z$$

Another way to work with these expressions is to stack them up and then add or subtract. This is especially easy to see if they give you parentheses.

$$(7a + 3b + 15c) + (2a + 4b + c) - (a - 2b + 2c)$$

If they are all addition, you can do them all at once, but here, because the last one is subtraction, we will just start with the first two expressions being added.

$$(7a + 3b + 15c)$$

$$(2a + 4b + c)$$

Once they are stacked with the like terms all in the same columns, you can add them.

$$\begin{array}{r} (7a + 3b + 15c) \\ +\ (2a + 4b + c) \\ \hline 9a + 7b + 16c \end{array}$$

Then, you can subtract the next bunch.

$$\begin{array}{r} 9a + 7b + 16c \\ -\ (a - 2b + 2c) \\ \hline 8a + 9b + 14c \end{array}$$

Hey wait a minute! Why do the b terms get subtracted that way?

Because you are handling the addition and subtraction the same way you would any positive or negative number. You are subtracting $-2b$ from $7b$.

$$7b-(-2b)$$

Remember when you subtract a negative number? You cancel the negative, so it's as though you add a positive.

$$7b+2b=9b$$

So your final result after adding and then subtracting is $8a+9b+14c$. Much more simplified than all those parentheses.

Try to combine and simplify a few expressions on your own.

Quiz #5

1. $4a+5a$
2. $3x-4y+2x+y$
3. $13b-5b+6b$
4. $23x-15x+x$
5. $(3a-3b)+(2a+2b)$
6. $22x-11x$
7. $(3x-2y)+(12x+14y)$
8. $15a-17a$

Now there are a few other definitions that will help you, because some people really like to call things by fancy names. When an expression has only one term in it, it is called a **monomial**. "Mono" means one, or alone. Like a monocle is that eye piece that covers only one eye that people wear to ballrooms in old movies.

$6xy$ is a monomial

When an expression has more than one term in it, it is called a **polynomial**. "Poly" means many, like a polyglot is someone who can speak many languages.

$$3ab+2b \text{ is a polynomial, and}$$

$$4x^2+3xy+y^2 \text{ is also a polynomial}$$

Polynomial covers all expressions of more than one term. But some people get off on making distinctions, so here are some even more specific ways to talk about these expressions.

If an expression has exactly two terms in it, it is called a **binomial**.

$$3x+7y \text{ is a binomial}$$

If an expression has exactly three terms in it, it is called a **trinomial**.

$$6a+4b+5c \text{ is a trinomial}$$

SIMPLIFYING BY MULTIPLYING AND DIVIDING EXPRESSIONS

Now that you know how to add and subtract to simplify expressions, you need to know how to multiply and divide expressions. It's just another step in simplifying and sometimes solving these expressions and equations. Here is where your understanding of the distributive property and everything you learned in the chapter about numbers is going to come in mighty handy. If you have any hesitation about that stuff, take a quick read back to page 31 and reacquaint yourself. The distributive property goes like this:

$$a(b+c)=ab+ac$$

Using numbers, $4\times5=20$ and $4\times3+4\times2=20$, because

$$4\times3=12, \text{ and } 4\times2=8$$

$$12+8=20 \text{ also.}$$

Remember? It's helped you before, it will help you again. The thing is, you can multiply and divide terms, both like and unlike, to form new terms. Look for a moment.

$$3x(y+2z) = 3xy + 6xz$$

So, multiplying terms just means using the distributive property.

$$3x \times y = 3xy, \text{ and } 3x \times 2z = 6xz$$

You multiply both the numbers and the variables.
Try a more complex multiplication problem.

$$3(2x+1) - (x+3) + 2(x-3)$$

First things first. This expression needs to be multiplied in parts and added in parts. Remember PEMDAS? It is easiest to multiply first.

$$3(2x+1) = 6x + 3$$

So you have $6x + 3 - (x+3) + 2(x-3)$.
Then, the next multiplication.

$$-(x+3) = -x - 3$$

So you have $6x + 3 - x - 3 + 2(x-3)$.

It's as though having the minus sign in front is telling you that you are multiplying the terms in the parentheses by negative 1, instead of 1. This is because you are subtracting both the x and the 3. Now, the last part.

$$2(x-3) = 2x - 6$$

So you have $6x + 3 - x - 3 + 2x - 6$.
Now you can combine like terms.

$$6x - x = 5x, \text{ and } 5x + 2x = 7x$$

$$3 - 3 = 0, \text{ and } 0 - 6 = -6$$

$7x - 6$ is the simplified form of the expression.

As long as you are careful of the signs and remember how to combine variables, you'll be fine simplifying expressions and multiplying and dividing variables.

Because it works the same for division.

$$\frac{8xz+6yz}{2z}$$

$$\frac{8xz}{2z}+\frac{6yz}{2z}$$

Remember what happens when you divide anything by itself?

It becomes 1, so here since you divide $8xz$ by $2z$, the $\frac{8}{2}$ becomes 4, and $\frac{z}{z}$ becomes 1. Since you then multiply the $4x$ by 1, you just have $4x$.

$$\frac{8x\not{z}}{2\not{z}}$$

Since both terms are over the $2z$, the same thing happens to the $6yz$.

$$\frac{\not{6}y\not{z}}{2\not{z}}=3y$$

So, combining both terms would give you:

$$4x+3y$$

An easy way to think of it is as reducing a fraction.

If z were 10, and $\frac{6yz}{2z}$ were $\frac{6y10}{2\times10}$, it would be easy enough to reduce. And that is what you are doing; you are putting the expressions into a simpler form.

APPROXIMATE THIS:
How long a line could the average lead pencil draw?

Now, try to multiply and divide some expressions on your own.

Quiz #6

1. $\dfrac{90x}{30x}$

2. $5x(3x+2)$

3. $\dfrac{20a+14b}{2b}$

4. $\dfrac{18x \times 6y^2 \times 4z}{18x \times 2y}$

5. $\dfrac{6wx}{3x}$

6. $\dfrac{35ab}{7b^2c}$

7. $\dfrac{12a^2 \times b}{4ab}$

8. $4(5a+2)$

What you have been doing so far is multiplying and dividing by monomials. For a change of pace, take a look at how to multiply polynomials.

Do you feel comfortable with these sorts of terrifying words yet? As long as you know what they mean, you're on top of things. If you get confused, take a look back and review, or run to the end of the chapter and look at the definitions and examples again.

$$(x+y)(x-4)$$

Here you have a friendly little set-up to multiply some polynomials. Use the distributive property again. Multiply both parts of the second polynomial by x, and then multiply them both by y.

$$(x+y)(x-4)$$

$$x(x-4)+y(x-4)$$

$$x^2-4x+xy-4y$$

What you're doing here is multiplying the terms in the second expression by the terms in the first expression. Notice how x multiplied by 4 in the second term becomes $-4x$. You have to remember to include the sign; here it is subtracted so it has to be subtracted in the simplified version, too. That's how you multiply polynomials.

Look at another one, an even bigger one.

$$(a+b+c)(2a+3b)$$

Multiply each term in the second by each term in the first. Again, just use your pal the distributive property.

$$a(2a+3b)+b(2a+3b)+c(2a+3b)=$$

$$2a^2+3ab+2ab+3b^2+2ac+3bc$$

Can you combine like terms and simplify it? Why not!

$$2a^2+5ab+3b^2+2ac+3bc$$

By the way, most algebraic expressions are supposed to be put in the order of their "highest powers" as well as that alphabetical issue discussed earlier, so you would probably rearrange things a bit.

$$2a^2+5ab+2ac+3b^2+3bc$$

Congratulations. You now know how to handle gigantic polynomial multiplication problems. Here's one thing that might make it easier, especially when you have to multiply binomials.

$$(x+y)^2$$

This is equal to

$$(x+y)(x+y)$$

You already know how to multiply this. You multiply both terms in the first expression by both terms in the second expression. That's plain old distributive property. You might also want to know a shorthand way of referring to this process with binomials. FOIL. FOIL stands for First, Outside, Inside, Last.

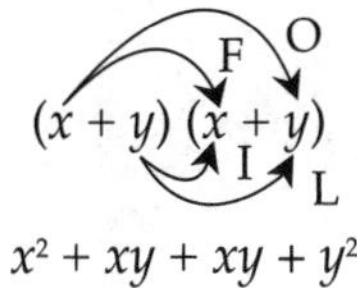

$$x^2 + xy + xy + y^2$$

You can combine like terms and simplify and write $x^2 + 2xy + y^2$.

This works when there are numbers in the expression, too.

$$(x+3y)(2x+y)$$

$$2x^2 + xy + 6xy + 3y^2$$

$$2x^2 + 7xy + 3y^2$$

It works out the exact same way as when you did it with the distributive property. In fact, you might want to use the distributive property on this expression just to prove it to yourself.

There are a few of these multiplied binomials that come up a lot on different standardized tests like the SAT and the GMAT and the GRE.

$$(x+y)(x+y) = x^2 + 2xy + y^2$$

$(x+y)(x-y) = x^2 - y^2$. This one is called the difference of two squares.

$$(x-y)(x-y)=x^2-2xy+y^2$$

Try working them out on your own. Check out that last one, multiplying $-y$ by $-y$ ends up giving you $+y^2$. In general, it's a good idea to know how to work these out, because understanding how it works makes you feel confident. But once you do know how, memorize those three you just looked at. It makes life a little easier, and that is worth a lot.

Try multiplying the following expressions:

Quiz #7

1. $(a+3b)(3a+b)$
2. $(ac-3)(c-b)$
3. $(4x-1)(y+1)$
4. $(w+z)(w-z)$
5. $(x+1)(x+1)$
6. $(e-f)(e-f)$
7. $(x+2)(3x+2)$
8. $\left(\frac{y}{2}+\frac{z}{3}\right)\left(\frac{y}{2}+\frac{z}{3}\right)$

Simplifying by Factoring

There is another way to simplify equations that goes back to the first chapter, when we talked about factors. If you don't feel sure about it, just check back to page 37. The factors of a number are numbers that, multiplied together, equal that number. For instance, the factors of 6 are (it's a good idea to list factors in pairs) 1,6 and 2,3.

You can also identify the factors of algebraic expressions. The way to factor expressions is to find a common factor of all the terms in an expression. Take a look at this one.

$$6x^2y+3xy^2$$

There are two terms in this expression, $6x^2y$ and $3xy^2$. The first thing you want to look at are the coefficients.

6 and 3.

Do these have any common factors? Sure, 3 is a factor common to both of them. You divide out the 3, and put the remaining terms in parentheses.

$$3\left(2x^2y + xy^2\right)$$

Now you want to check the variables for their factors.

x^2 and x. Factored out, $x^2 = x \times x$, and x is, of course, just x.

Well, x is a common factor to both of these, so now the common factor is $3x$. Divide it out and put it outside the parentheses.

$$3x\left(2xy + y^2\right)$$

Check the next variable.

y and y^2. You factor y and y^2, and get y, and $y \times y$. So, for both these variables, y is a common factor, and that means the common factor of your expression is now $3xy$. You write your common factor outside of the parentheses in which your simplified expression will now sit.

$$3xy\ (2x + y)$$

And that is what the expression looks like once it is simplified by factoring.

If you multiply it out again, you will see that you have the same expression as when you started.

$$3xy(2x + y) = 6x^2y + 3xy^2$$

The way to get that simplified expression is to figure the common factors, and then divide them out of the terms.

The common factor of an expression goes outside the parentheses, because you end up multiplying every term by that factor.

Try factoring another expression.

$$3a^2x + 9x^2 + 15x$$

Look at the coefficients first. 3, 9, and 15. What is their common factor? One way to do this is to factor each of them to their prime factors.

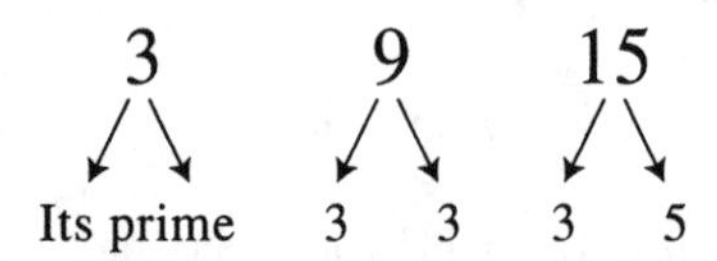

The common factor is 3. Put it outside the parentheses.

$$3\left(a^2x + 3x^2 + 5x\right)$$

Now look at the terms in parentheses, a^2x, x, and x^2. You can reduce them to their factors. The factors of a^2x are a, a and x. And the factors of x^2 are x and x. The common factor of these two terms is x.

$$3x\left(a^2 + 3x + 5\right)$$

It will help a lot to remember those expressions that were reviewed earlier, on page 179, as the popular expressions on the SAT, GMAT, and GRE.

$$(x + y)(x + y) = x^2 + 2xy + y^2$$

$$(x + y)(x - y) = x^2 - y^2$$

$$(x - y)(x - y) = x^2 - 2xy + y^2$$

You will run into these forms a lot, and they won't always be composed of just variables. For instance, take a look at this:

$$x^2 + 8x + 16$$

When you are asked to factor a trinomial like this, take a look at the last term. What are its possible factors?

1,16   2,8   4,4

One of these pairs of factors will combine by addition or subtraction to form the coefficient in the second term. In this case, which pair of factors will add up to 8? 4,4. This means the factored form of this expression is

$$(x+4)(x+4) \text{ or, } (x+4)^2$$

To check it, multiply it out.

$$(x+4)(x+4) = x^2 + 8x + 16$$

This will work even if you have two different numbers in your expression.

$$a^2 + 2a - 15$$

What are the possible factors of 15? 1,15 and 3,5.

Which factor pair combines in some way to form 2? Definitely not 1 and 15. If you add 1 and 15, you get 16, and if you subtract you get either 14 or -14. So the workable pair must be 3 and 5. The trick is that in this case, you need to subtract. And to get positive 2—which you have in the expression, $+2a$—you need to subtract 3 from 5. So the 5 must be in the part of the expression being added, and the 3 must be in the part being subtracted. This means a subtracted 3 and an added 5. How do you know one expression has subtraction? Because the 15 in the original expression is being *subtracted*. You're looking for factors of negative 15. So your expression looks like this:

$$(a+5)(a-3)$$

Multiply it out to check your work.

$$(a+5)(a-3) = a^2 + 2a - 15$$

The factors of this expression are $(a+5)$ and $(a-3)$.

Look at one more, and keep an eye out for where the minus sign appears in each of these examples, and how it appears when the expression has been factored.

$$r^2 - 9r + 20$$

As you have been doing with all the other trinomials, you want to take a look at the third term. What are its possible factors? 1,20 and 2,10 and 4,5. Which of these pairs combine to form –9? 4,5. But to combine to –9, you will need to have both the 4 and the 5 be subtracted.

$$(r-4)(r-5)$$

Now you can check your work, and see that when you have the two negative factors multiplied, you will get a positive 20.

Look back at the way these work. It's important to pay attention to the sign in the trinomial being factored. Here are some useful generalizations:

1. When you have a pair of factors both featuring subtraction, the second term of the trinomial is being subtracted, and the third term is being added.

$$(r-4)(r-5) = r^2 - 9r + 20$$

2. When the pair of factors has one set of addition, and another of subtraction, the third term of the trinomial is subtracted.

$$(a+5)(a-3) = a^2 + 2a - 15$$

3. When the two pairs are both being added, the third term of the trinomial is being added, and the second term is being added as well.

$$(x+4)(x+4) = x^2 + 8x + 16$$

One of the ways these fancy polynomials will be used is to give possible values for the variable in the expression. These are called **roots** of the equation. Take a look at this.

$$x^2 + 7x + 12 = 0$$

What's different here? The expression has been set equal to 0 to form an equation. What does this mean when you factor it?

$$(x+3)(x+4)=0$$

This means that these two factors give the product 0. Now you know that in order to get a product of 0, you need to have one of the factors be 0. Either $x+3=0$ or $x+4=0$. That means that for this example, x could equal one of two numbers, -3 or -4. That means -3 and -4 are the roots of this equation. Try it out.

If x is equal to -3, then $(-3+3)(-3+4)=0$.

If x is equal to -4, then $(-4+3)(-4+4)=0$.

You don't know the exact value of x in this case, but you know it must be either -3 or -4.

Try factoring and simplifying the expressions that follow. If they are equations, give possible values for the variables.

Quiz #8

1. $2w^2xy^2-8wy$

2. $x^2-8x+12=0$

3. $15ab+3a^2$

4. $c^2+9c+18=0$

5. $a^2+10a+25=0$

6. $8a^2b-16b^2$

7. $\dfrac{x^2-25}{x+5}=0$

8. $\dfrac{n^2+5n-14}{n-2}=0$

A TEENY BIT MORE FACTORING

One of the ways in which factoring can throw a curve ball at you is by having either the first term or the third term in the trinomial be complex.

For example: $4x^2 + 12x + 8$.

Look at that first term. It has a coefficient that isn't equal to 1 the way the other expressions we have looked at have been. What to do here is to look at that first coefficient and figure out its factors. It could be 1,4 or 2,2. You will probably need to try them both. Look at the third term, as you did before, to figure its factors.

$(4x+2)(x+4) = 4x^2 + 2x + 16x + 8$. No good. Try again.

$(2x+2)(2x+4) = 4x^2 + 4x + 8x + 8 = 4x^2 + 12x + 8$. You've hit it.

Unfortunately, the way to factor these kinds of expressions is through trial and error, with a bit of help in knowing the factors. Fortunately, they don't show up much on the SAT, the GRE, or the GMAT.

> **APPROXIMATE THIS:**
> How much food and drink does the average person consume in one year?

SIMPLIFYING BY FRACTIONS

You'll come across another way of simplifying algebraic expressions when you are asked to change algebraic expressions to fractions. When this happens, all you have to remember is that the way to handle fractions in algebra is the exact same way you handle fractions with any numbers at all. This means things like common denominators, and multiplying by an expression over itself, which is still going to equal 1, all these ideas are still highly useful. You are using the bow tie, pages 66 and 82 or, again. Take a look at an example:

$$3x + \frac{x+2}{2x}$$

Part of the expression is not in fraction form, $3x$, and the other part is, $\frac{x+2}{2x}$. If someone asked you to put this into fraction form, you would want the whole thing to have a common denominator. You are adding fractions, just like the good old days of chapter 2. Turn back to page 82 if you want to review. Since you already have a denominator of $2x$, use that. It is a bit like a mixed number this way. How will you get a denominator of $2x$ on your friend $3x$ over there? By using the bow tie.

$$\overset{6x^2}{\searrow}\quad \frac{3x}{1} + \frac{x+2}{2x} =$$

$$\frac{6x^2}{2x} + \frac{x+2}{2x} = \frac{6x^2 + x + 2}{2x}$$

It makes sense, doesn't it. Take a look at another one.

$$a + 2 + \frac{3a+6}{a+2}$$

Well, you've got a scary-looking denominator in the form of $a+2$. But you're feeling brave (we hope), so you go ahead without a second thought and use the bow tie.

$$\frac{a+2}{1} + \frac{3a+6}{a+2}$$

Here's where it might help to have memorized those frequently occurring multiplying polynomials before. Remember $(x+y)(x+y)$? Well $(a+2)(a+2)$ is the same thing really, with a as x, and 2 as y.

$$\frac{a^2 + 4a + 4 + 3a + 6}{a+2}$$

Now you can combine like terms.

$$\frac{a^2+7a+10}{a+2}$$

There you go, there's your fraction. You can simplify it even more, can't you? To simplify it all the way, you can factor. What are the possible factors of the third term, 10? 1,10 and 2,5. Which pair will combine to form 7?

$$\frac{(a+5)(a+2)}{(a+2)}$$

What is $\frac{a+2}{a+2}$? Any number over itself is equal to 1. So the expression $(a+2)$ cancels, and there you have your simplified expression.

$$(a+5)$$

Algebra follows those same rules you've learned with ordinary math, so don't get terrified when you're asked to do something you know how to do in regular math, just follow the same steps and you will be fine.

Try to express the following in fraction form, and don't forget to simplify it once you do.

Quiz #9

1. $\frac{x+4}{x}+(x+y)$

2. $w+4-\frac{w+4}{w-4}$

3. $a-\frac{a+b}{2}$

4. $\frac{r-s}{t-s}+t+s$

5. $\frac{a}{s}+\frac{r}{t}$

6. $\frac{p+q}{p-q}+\frac{p-q}{p+q}$

7. $5c-\frac{c+d}{c}$

8. $x+1+\frac{x^2+x-3}{x}$

Excellent job. Now you may be asking, "Why is all of this about putting expressions into a million different forms, but never solving them?"

It's because being able to manipulate equations is more than half the battle with algebra. Most of the time on tests, all you will be asked to do is manipulate equations. It is a way of showing that you understand what the pieces mean when they are arranged in abstract form. Once you start to feel comfortable with moving things around, you've conquered vast areas. And once you have conquered vast areas, you should feel pretty good about all this, maybe even extremely math smart. You have already seen some ways to use algebra to find number solutions for equations like substituting in numbers for variables. That method works when you have an equation with one variable in it, like $3g=27$. What about equations that have two variables? Now we can talk about another way of finding solutions.

SIMULTANEOUS EQUATIONS

If you have an equation with two variables, like $a+b=6$ you can't really solve for either variable in this form. The two variables could be 1 and 5, or –3 and 9, or $5\frac{1}{2}$ and $\frac{1}{2}$. They could be anything. And, even if you knew which pair of numbers it was, you wouldn't know which variable represented which number. But, if you had another equation using the same variables in a different combination, you could solve it. It's called **simultaneous equations** because you are using two equations at the same time. What if you had another equation that said $2a+b=10$? Look at them together.

$$a + b = 6$$

$$2a + b = 10$$

One of the ways to work with simultaneous equations is to add or subtract them from one another to get rid of one variable. For example, if you wanted to find what a stood for, you would eliminate the b variable. You can put the bigger equation on top, and then subtract them just as if they were numbers.

$$\begin{array}{r} 2a + b = 10 \\ \underline{- \; a + b = \;\; 6} \\ a + 0 = \;\; 4 \end{array}$$

To find b, substitute a back into one of the expressions.

$$4 + b = 6$$

$$b = 2$$

Using two equations containing the same two variables in different configurations, you can determine the values of both variables.

Now look at an example where you need to add the equations:

$$x - 3y = 9$$

$$2x + 3y = 36$$

To get a variable alone, it will be easy here to add the two equations because then the two $3y$ terms will cancel one another out.

$$\begin{array}{r} x - 3y = \;\; 9 \\ \underline{+ \; 2x + 3y = 36} \\ 3x + 0 \;\; = 45 \\ x \qquad = 15 \end{array}$$

Now you can substitute the value for x back into either of the equations.

$$30 + 3y = 36$$
$$3y = 6$$
$$y = 2$$

Why This Works in Such a Cool Way

When you are given two equations with the same variables, like $a + b = 6$ and $2a + b = 10$, the equations are saying, "Here are a few things that are true. If you combine a and b, you get 6, and if you combine $2a$ and b, you get 10." That means $a + b$ is actually a number, and that number is 6. Now you can add 6 to other numbers, can't you? That means you can add $a + b$ to other expressions.

If you added the same expressions, you would have a tough time finding anything new.

"Well, $a + b = 6$, and $a + b = 6$, which tells me . . . " But if you have two different pieces of information, there is only one way for them to intersect, or both be true.

Try another set, one that is a bit different.

$$2x + 3y = 7$$

$$3x + 2y = 8$$

You look and add them and you don't get rid of any variable, you subtract and you don't get rid of any variable, so what can you do? What you can do is multiply one or both of the equations so one set of like terms will have the same coefficient. Sounds fancy but it just means you want it to look like the other pairs of equations looked. Let's try to make the terms with x have equal coefficients. An easy way to do this is to use the coefficients of these x terms as factors. It's a bit like finding a common denominator.

$$3(2x + 3y = 7)$$
$$2(3x + 2y = 8)$$

Why does the entire equation get multiplied? Because remember the rule of algebra, anything you do is okay, as long as you do it to all parts and both sides of the equation.

$$6x + 9y = 21$$
$$6x + 4y = 16$$

Now you can subtract, same as you did before.

$$\begin{array}{r} 6x + 9y = 21 \\ -\ 6x + 4y = 16 \\ \hline 0 + 5y = \ \ 5 \end{array}$$

You need to isolate *y*.

$$y = 1$$

Now you can substitute *y* back into either one of the equations.

$$2x + 3 = 7$$
$$2x = 4$$
$$x = 2$$

APPROXIMATE THIS:

What is the ratio of your height to the measurement from fingertip to fingertip of your arms and chest, measured across?

Try solving a few more on your own. Solve for both variables.

Quiz #10

1. $x + y = 3$ and $x + 2y = 4$
2. $2v + w = 13$ and $v + 3w = 19$
3. $a + b = 12$ and $a - b = 4$
4. $3c + 2d = 22$ and $2c - 2d = 8$

5. $e - 5f = 0$ and $e - 6f = -2$
6. $4r + 2s = 20$ and $r + 3s = 10$
7. $a - b = 7$ and $a - 4b = -2$
8. $m + n = -6$ and $m - n = -2$

INEQUALITIES

This leaves you another area of algebra to tackle. **Inequalities.**

$$x > 2$$

Inequalities can be used with either numbers or variables or both, and they are a bit more vague than other parts of math, because x could be 2.000001 or x could be 1,000,000. The inequality here means "x is greater than 2."

They are used to express the idea that the one side of an algebra problem is either bigger or smaller than the other. This means you don't know exactly what x is equal to; it could be 3, it could be 10,000,000, it could be 2.0000000000001. All you know about x is that it is bigger than 2. The easiest way to read inequalities is to look at which side of the inequality sign is bigger.

$$x > 2$$

Inequality signs also show up in the other direction.

$$a < 5$$

All you have to do is look where the bigger side or the open side of the inequality sign is. It's on the 5, right? So 5 is bigger than a. The other possibilities of jazzing up the sign are these.

$$x \geq 2$$
$$a \leq 5$$

These signs mean you have one more option, x is greater than or equal to 2. That means x could be any one of the infinite numbers bigger than 2, AND it could also equal 2. And a could be any one of the infinite numbers less than 5, AND it could be equal to 5.

So you are probably asking yourself, "And in what wonderful ways will I be able to apply this new information to algebra?" Well, let's take a trip back to manipulating equations.

Let's say I said that even if I had $100 less than I do, I would still have more than $500. How would you write that out in algebra?

Set up money as M, and then write an inequality.

$$M - 100 > 500$$

What that says is, "the money minus one hundred is greater than five hundred."

Now you want to isolate your variable, and you approach it in the exact way you approached isolating variables when you were working with equations (things that are equal). That means, to get the M by itself, you need to perform the reverse operation. One hundred is being subtracted, so you need to add one hundred, and you need to add it to both sides of the inequality.

$$M - 100 + 100 > 500 + 100$$

$$M > 600$$

This means that I have more than $600.

What if the variable hadn't been first? How do you manipulate the equation in that situation?

$$30 - b > 10$$

If you just subtract 30 from both sides, you will get negatives and for right now you don't want to deal with that. What you can do, as you can with any algebraic expression, is move the variable around, too. To perform the reverse operation, you will have to add *b*, to both sides of course.

$$30 - b + b > 10 + b$$

$$30 > 10 + b$$

To isolate the variable, which of course you still need to do, you want to get rid of the 10, but it is much less traumatic this way.

$$30 - 10 > 10 + b - 10$$

$$20 > b$$

Try a few more solo:

Quiz #11

1. $b - 3 > 7$
2. $16 - c < 4$
3. $10 + x \geq 19$
4. $y - 7 \leq 10$
5. $33 - a > 23$
6. $13 + x > 15$
7. $x - 6 \leq 12$
8. $y + 2 \geq 14$

You know how to handle adding and subtracting numbers in your manipulation of inequalities, so think about multiplying and dividing.

$$4x > 12$$

To isolate the variable here, the reverse operation is division.

$$\frac{4x}{4} > \frac{12}{4}$$

$$x > 3$$

No big deal. But what about when you have a negative coefficient being multiplied or divided? Then the operation gets one more step.

$$-5y < 20$$

To isolate, you need to divide. But when you divide or multiply by a negative number in an inequality, you turn the inequality sign around to face the other way. What was the bigger side becomes the smaller side.

$$\frac{-5y}{-5} > \frac{20}{-5}$$

$$y > -4$$

If you want to check your work, you can plug a value for the variable back into the inequality and see if it works. Since $y > -4$, try -3. Remember your number line if you've forgotten how to determine which number is bigger.

$$-5 \times -3 < 20$$

$15 < 20$. It's true, right?

You can also solve inequalities in terms of variables, just like equations.

If $y + 3x < 2x + 6$, express x in terms of y.

This means you need to isolate x. Try subtracting $2x$ from both sides.

$$y + 3x - 2x < 2x + 6 - 2x$$

$$y + x < 6$$

Now, put y on the other side by subtracting it.

$$x < 6 - y$$

This expresses x in terms of y.

Try manipulating a few more inequalities:

Quiz #12

1. $x - 5 > 3x + 1$
2. $c + 4 < -5$
3. $16 - y > 4$
4. $12a + 2 \leq 34$
5. $13 + \frac{n}{-3} \geq 33$
6. $7 \leq v - 10$
7. $10 + x < 14 - 3x$
8. $4x - 5 \geq 2x + 1$

Once you feel comfortable with the algebra just discussed in this chapter, you are prepared for any of the algebra that appears on the big standardized tests. Which should feel pretty comforting.

GLOSSARY

algebra: Math that uses letters to represent numbers.
variable: A letter used to represent a number.
equation: Two expressions set equal to each other.
formula: An expression representing a set relationship among the terms.
coefficient: The number multiplied by a variable in a term.
unknown: A variable.
expression: Terms combined by addition or subtraction.
term: A number or variable or both combined by multiplication or division.
like terms: Terms containing the same variables.
monomial: An expression containing one term.
polynomial: An expression containing more than one term.
binomial: An expression containing exactly two terms.
trinomial: An expression containing exactly three terms.
root: NOT the square root. A possible solution for a factored expression.
simultaneous equations: Separate binomial equations in which each term has only one variable. Both equations contain the same two distinct (different) variables.
inequality: An expression where one side is set greater than or both greater than or equal to another.

To make sure you are okay with it, try these questions that combine all the different algebraic stuff covered in this chapter.

Quiz #13

1. Joan is 4 years older than her boyfriend, Bob, who is 2 years younger than Carol, who is 17 years old. How old is Joan?

2. Put $v+3+\frac{2v-4}{v}$ into simplified fraction form.

3. Fred has to get his parents' car back to their house by 5 o'clock. If it is now 3 o'clock and his parents' house is 240 miles away, at what rate does Fred have to drive? (Remember $R \times T = D$.)

4. In the equation $x^2 - 15x + 56 = 0$, what are the possible values of x?

5. If $x + 3y = 25$, and $2x = y + 1$, what are x and y?

6. In the inequality $13x + 5 < -5x - 2$, isolate x.

7. If Susan has 5 more than twice the number of guitars that Karen has, and Karen has 6 guitars, how many guitars does Susan have?

8. Factor the equation $x^2 + xy = -y^2 - xy$.

9. In the equation $x^2 - 2x = 15$, what are the possible values for x?

10. In the inequality $-5x + 2 < 3y + 6$, solve for y in terms of x.

ANSWER KEY

QUIZ #1

1. $G = \frac{B}{2}$
   G is girls, B is boys.

2. $T = \frac{8}{100P}$
   T is tax, P is price.

3. $S = 4H$
   S is shoes, H is hat.

4. $S = M - 4$
   S is Stanley, M is Margaret.

5. $C = 3T + 4L$
   C is computer, T is typewriter, and L is calculator. You can't use C again, because then you would have C representing two different things in the equation, and remember, within an equation a variable only represents one thing.

6. $G = \frac{B}{3}$
   G is George's weight, B is Ben's weight.

7. $D = H + 3$
   D is David, H is Hanna.

8. $B = 3L$
   B is books, and L is library visits. If you had trouble with this one, ask yourself which do I end up with more of, books or library visits? Books, right? Because for every one library visit you get 3 books, so to have the equation balance out to be equal, you need to put the number, or coefficient, next to the library.

QUIZ #2

1. $x + 4 = 7,\ x = 3$
2. $2x = x - 3,\ x = -3$
3. $x + 6 = 15,\ x = 9$
4. $3x = x + 14,\ x = 7$
5. $x + 3 = 40 - x,\ x = 18.5$
6. $\frac{x}{8} = 6,\ x = 48$
7. $40 - x = 3x,\ x = 10$
8. $x + 9 = 2x - 3,\ x = 12$

QUIZ #3

1. $a = \frac{4 + b}{3}$
2. $y = x$
3. $z = w + 4$
4. $e = \frac{f}{3}$
5. $b = a + 13$

6. $x = 5 - y$, you can also say $x = -y + 5$; it's the same thing.

7. $b = \dfrac{a-2}{3}$

8. $c = 4 + d$

Quiz #4

1. Two terms, $7x$ and $6xy$; 7 is the first coefficient, and 6 is the second coefficient.
2. There are three terms, $-4x^2$, $3y$, and 5; 4 is the first coefficient, and 3 is the second coefficient.
3. There are three terms, $7a$, $3ab$, and b^2; 7 is the first coefficient, 3 is the second coefficient, and 1 is the invisible third coefficient.
4. Three terms, $3c$, b^3, and 7. The first coefficient is 3, and the second coefficient is the invisible 1.
5. Two terms, $4x$ and $4y$; both coefficients are 4.
6. Two terms, $3a$ and $2b$; the first coefficient is 3 and the second coefficient is 2.
7. Three terms, $4a^2$, $2ab$, and b^2. The first coefficient is 4, the second coefficient is 2, and the third coefficient is the same old invisible 1.
8. Two terms, $2x$ and $5y$. The first coefficient is 2 and the second coefficient is 5.

Quiz #5

1. $9a$
2. $5x - 3y$
3. $14b$
4. $9x$
5. $5a - b$
6. $11x$
7. $15x + 12y$
8. $-2a$

Quiz #6

1. 3
2. $15x^2 + 10x$
3. $\dfrac{10a + 7b}{b}$
4. $12yz$
5. $2w$
6. $\dfrac{5a}{bc}$
7. $3a$
8. $20a + 8$

Quiz #7

1. $3a^2 + 10ab + 3b^2$
2. $ac^2 - acb - 3c + 3b$
3. $4xy + 4x - y - 1$
4. $w^2 - z^2$
5. $x^2 + 2x + 1$
6. $e^2 - 2ef + f^2$
7. $3x^2 + 8x + 4$
8. $\dfrac{y^2}{4} + \dfrac{y^z}{3} + \dfrac{z^2}{9}$

Quiz #8

1. $2wy(wxy-4)$
2. $(x-6)(x-2)=0$, $x=6$ or 2
3. $3a(a+5b)$ You switch the items in the parentheses for alphabetical reasons.
4. $(c+6)(c+3)=0$, $c=-6$ or -3
5. $(a+5)(a+5)=0$, $a=-5$
6. $8b(a^2-2b)$
7. $\dfrac{(x+5)(x-5)}{(x+5)}=0$

   $(x-5)=0$, $x=5$
8. $\dfrac{(n-2)(n+7)}{(n-2)}=0$

   $(n+7)=0$, $n=-7$

Quiz #9

1. $\dfrac{x^2+xy+x+4}{x}$
2. $\dfrac{w^2-w-20}{w-4}$
3. $\dfrac{a-b}{2}$
4. $\dfrac{r-s^2-s+t^2}{t-s}$
5. $\dfrac{at+rs}{st}$
6. $\dfrac{2(p^2+q^2)}{(p^2-q^2)}$
7. $\dfrac{5c^2-c-d}{c}$
8. $\dfrac{2x^2-2x-3}{x}$

Quiz #10

1. $y=1$, and $x=2$
2. $w=5$, and $v=4$
3. $a=8$, and $b=4$
4. $c=6$, and $d=2$
5. $e=10$, and $f=2$
6. $r=4$, and $s=2$
7. $a=10$, and $b=3$
8. $m=-4$, and $n=-2$

Quiz #11

1. $b>10$
2. $12<c$
3. $x\geq 9$
4. $y\leq 17$
5. $10>a$
6. $x>2$
7. $x\leq 18$
8. $y\geq 12$

Quiz #12

1. $x<-3$
2. $c<-9$
3. $12>y$
4. $a\leq\dfrac{8}{3}$
5. $n\leq -60$
6. $17\leq v$
7. $x<1$
8. $x\geq 3$

QUIZ #13

1. Joan is 19.
   How It Works:
   Set it up as an equation. Joan is 4 years older than Bob, first.
   $$J = B + 4$$
   Then, Bob is 2 years younger than Carol.
   $$B = C - 2$$
   Then substitute in the 17 for Carol.
   $B = 17 - 2$, so $B = 15$.
   Then substitute in the 15 for B.
   $$J = 15 + 4, \quad J = 19.$$

2. $$\frac{v^2 + 5v - 4}{v}$$
   How It Works:
   You need a common denominator, so use the only denominator you have, v. Multiply $v + 3$ by 1 in the form of $\frac{v}{v}$.
   $$\frac{v}{v}(v + 3) = \frac{v^2 + 3v}{v}$$
   Now add it to the fraction you already have.
   $$\frac{v^2 + 3v}{v} + \frac{2v - 4}{v} =$$
   $$\frac{v^2 + 3v + 2v - 4}{v}$$
   Combine like terms.
   $$\frac{v^2 + 5v - 4}{v}$$

3. Fred has to drive 120 mph to get back in time.
   How It Works:
   You get to substitute numbers into a formula, yippee. $RT = D$. Well R is rate, and that is what is unknown, T is time, 2 hours to get back, and D is distance or 240 miles. $R\,2 = 240$. To solve for R, isolate your variable by performing the reverse operation, in this case, division.
   $$R \times \frac{2}{2} = \frac{240}{2}, \quad R = 120$$

4. x is either 7 or 8.
   How It Works:
   Factor factor factor! What are the possible factors of 56? 1,56, and 2,28, and 4,14, and 7,8. Well, which of these pairs could combine to 15? 7 and 8 of course. And you have a minus sign in front of the second terms, and a plus sign in front of the third term, it is a subtraction times a subtraction. $(x - 7)(x - 8) = 0$. Multiply it out to check it. Now how could these factors multiply to 0? One or the other of the pairs is equal to 0, so either $x - 7 = 0$, so $x = 7$, or $x - 8 = 0$, and $x = 8$.

5. $x = 4$, and $y = 7$.

How It Works:

Simultaneous equations can make your life much easier. To find *x*, first line up the equations. $x + 3y = 25$, and $2x - y = 1$. See how you can manipulate that second equation by subtracting *y* from both sides, so it looks like the first equation? Now line them up.

$$x + 3y = 25$$
$$2x - y = 1$$

To isolate *x*, you need to make the *y* terms cancel, so make them have the same coefficient, in this case 3.

$$x + 3y = 25$$

3(2*x* – *y* = 1 becomes 6*x* – 3*y* = 3. Now you can add them to remove the *y* terms.

$$x + 3y = 25$$
$$+ (6x - 3y = 3)$$

$7x = 28$, and divide both sides by 7 to isolate the *x*. $x = 4$. Now you can substitute the 4 back in for the *x*, as in $4 + 3y = 25$. $3y = 21$, $y = 7$. Or you can add the two equations to get rid of the *x* terms.

$$2(x + 3y = 25)$$

$2x - y = 1$ becomes

$$2x + 6y = 50$$
$$-(2x - y = 1)$$

$7y = 49$, $y = 7$. Either way, it makes good sense, right?

6. $x < \frac{-7}{18}$

How It Works:

You want to put all the numbers on one side, and the variables on the other. First, tip the inequality so the variables go to where there are already variables. $13x + 5 < -5x - 2$. There are 13*x* terms on the left side, and –5*x* terms on the right. You might as well put them all on the left by adding 5 *x* terms to both sides.

$18x + 5 < -2$. Now, move the 5 to the side with the numbers, the right side. You need to perform the reverse operation, so you subtract.

$18x < -7$. Now, to isolate the *x*, the reverse operation is division.

$x < \frac{-7}{18}$. Not too shabby.

7. Susan has 17 guitars.
How It Works:
You need to write an equation for the English here. Susan has 5 more than twice what Karen has. So Susan is S, Karen is K. Five more is +5, and twice means times 2.

$$S = 5 + 2K$$

Susan has, meaning equals or is, five more or +5 than twice the number Karen has, or $2K$. Now you can substitute in the value you are given for Karen, or 6 guitars.
$S = 5 + 2 \times 6$, better known as $S = 5 + 12$, Susan has 17 guitars.

8. $(x + y)^2 = 0$
How It Works:
When you are factoring an equation it is easiest to have all the variables on one side. In this case if you put all the variables on the right side by adding y^2 and xy to both sides, you get $x^2 + 2xy + y^2 = 0$.
This probably looks familiar by now.

$$(x + y)(x + y) = 0$$

or $(x + y)^2 = 0$.

9. $x = -3$ or 5
How It Works:
Another factoring situation. Again, when you have an equation, especially one with an x^2 in it, you probably will have an easier time of things if it is all on one side of the equation and set equal to 0. So subtract 15 from both sides of the equation.

$$x^2 - 2x - 15 = 0$$

Your third term is 15, and it is being subtracted. And your second term is being subtracted. That means one of your factors is being subtracted, and the other is being added. And what are the possible factors of 15 anyway? 1,15 and 3,5. Which will combine to give you 2? 3,5. And you want the 2 subtracted, so you need to have an added 3 and a subtracted 5.
$(x + 3)(x - 5) = 0$. To have these factors multiply to be 0, one of them has to equal 0, so either $x + 3 = 0$, and $x = -3$, or $x - 5 = 0$, and $x = 5$.

10. $y > \frac{-5x-4}{3}$

How It Works:

You want y by itself, and all numbers and x variables on the other side of the inequality. So first move the 6 by subtracting it from both sides.

$-5x + 2 - 6 < 3y + 6 - 6$

$-5x - 4 < 3y$. Now, the 3 needs to be moved, and the reverse operation is division. $\frac{-5x-4}{3} < y$. This is fine, but to make it all pretty, and still have the same information, just turn it around as it is shown in the correct answer. It means the same thing; y is bigger than $\frac{-5x-4}{3}$, it just shows it with y appearing first. Convention and style and all that, you understand. The y term is running the show at this point, because you are expressing the whole inequality "in terms of y" so it is only polite to have it first, in a way.

CHAPTER 6

GEOMETRY

Geometry is the branch of math that deals with shapes and space. The first things to understand in geometry are points and lines. For these you will use **graphs**, also known as **Cartesian grids**.

Remember the number line? Well graphs are nothing more than two number lines, crossed.

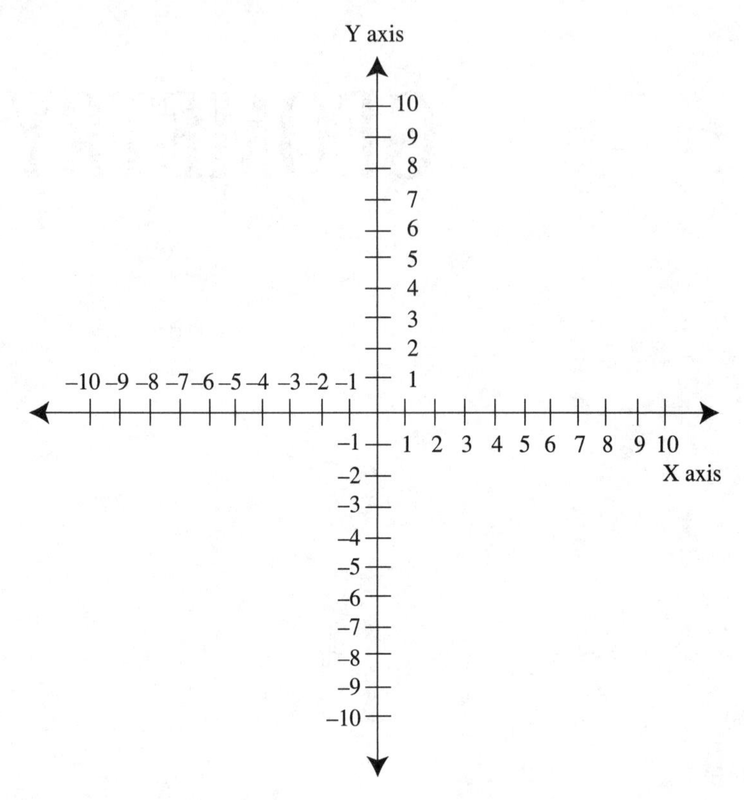

The crossing of the lines separates the space on the page into four sections. They have special little names. Since there are four spaces, they are called **quadrants**. "Quad" is the prefix meaning four. The individual quadrants' names are actually numbers, a big surprise, right? and they are numbered counterclockwise.

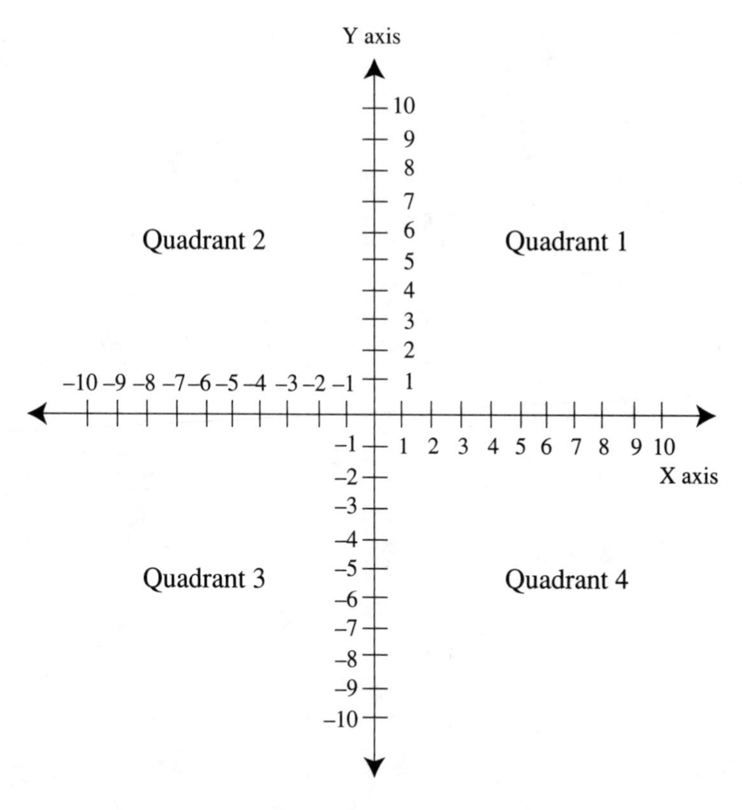

The horizontal number line is called the **X axis**, and the vertical number line is the **Y axis**. An axis is a straight line.

The place where the two number lines, or **axes** (pronounced ***axe-eez***, that's plural for axis), cross is called the **origin**. The origin point is the point 0,0. That means it is at the 0 point on the X axis, and at the 0 point on the Y axis. Every point on a graph is assigned two numbers. If you wanted to mark on the graph $x = 5$, and $y = 0$, you would count along the X axis 5 points from the origin. To mark the $y = 0$, you stay along the X axis because since you start at the origin, y already equals 0.

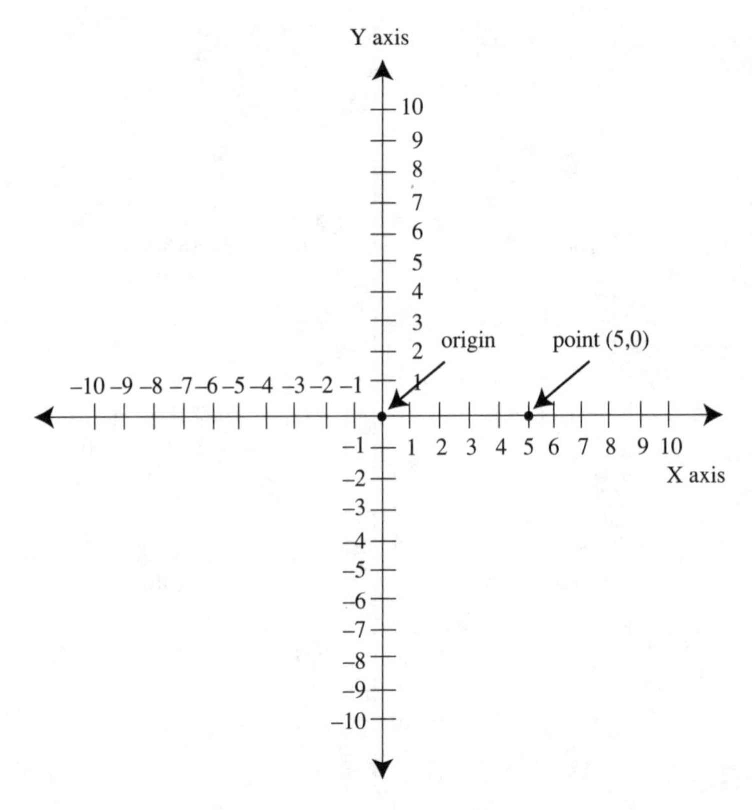

That is called **plotting a point**. A point takes up no space either vertically or horizontally, it's just a point. This particular point is called (5,0) because when you name points, you always give the *x* point first, probably because mathematicians are so into alphabetizing everything. The numbers (5,0) are called the **coordinates** of the point. And they get shown in parentheses because, well just because that's the way they look best.

As you can see, on the X axis, everything to the right of the origin is positive, and everything to the left of the origin is negative. For the Y axis, everything higher than the origin is positive, and everything lower than the origin is negative. This gives you a bit more information about your quadrants. Just like the number line.

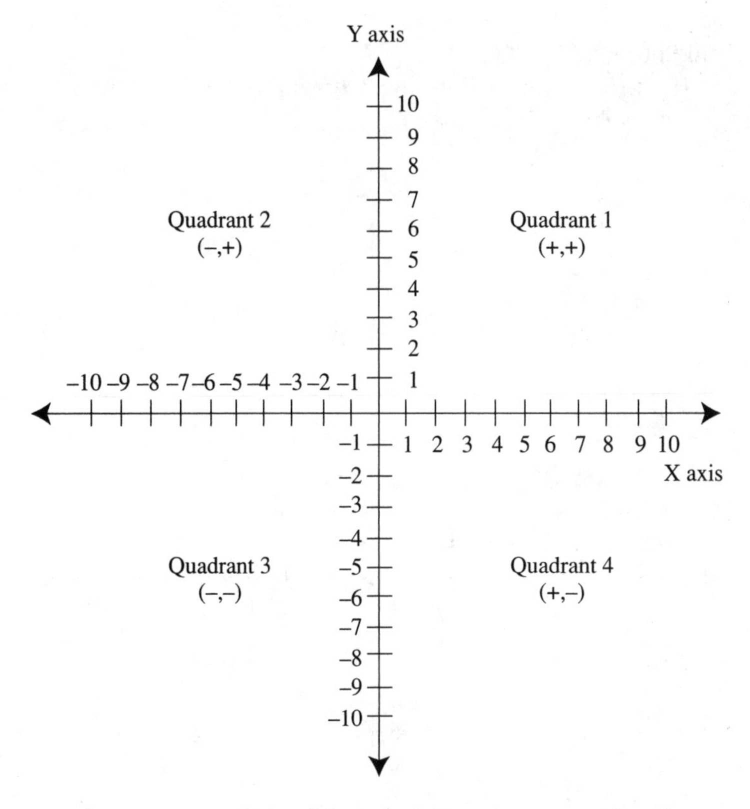

- If a point is in the first quadrant, the x coordinate is positive (right of the origin), and the y coordinate is positive (above the origin).

- If a point is in the second quadrant, the x coordinate is negative (left of the origin), and the y coordinate is positive (above the origin).

- If a point is in the third quadrant, the x coordinate is negative (left of the origin), and the y coordinate is negative (below the origin).

- If a point is in the fourth quadrant, the x coordinate is positive (right of the origin), and the y coordinate is negative (below the origin).

APPROXIMATE THIS:

How many times does the average human heart beat in 24 hours?

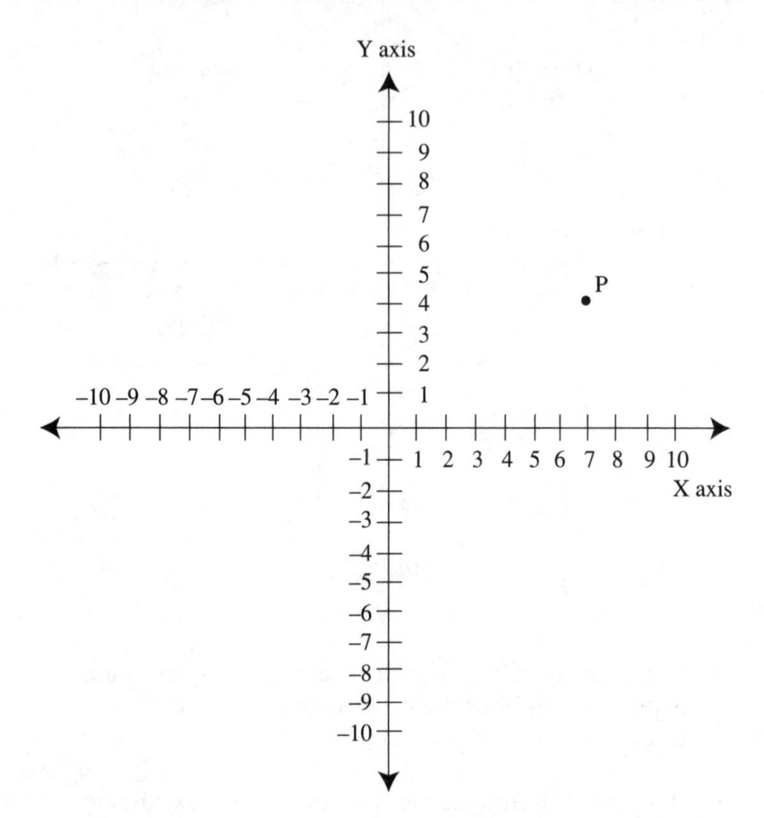

Look at point P. What are its coordinates? Well, count how far from the origin the x coordinate is by starting at the origin and counting along the X axis. It is seven spaces to the *right*, so the x coordinate is positive 7. To find the y coordinate, count along the Y axis to see how many spaces from the origin it is. It is four spaces *up*, so the y coordinate is positive 4. The coordinates of point P, known among some sophisticated folk as **ordered pairs**, are (7,4).

Quiz #1

Name the coordinates, and the quadrants, of the points ABCDEFGH labeled on the graph.

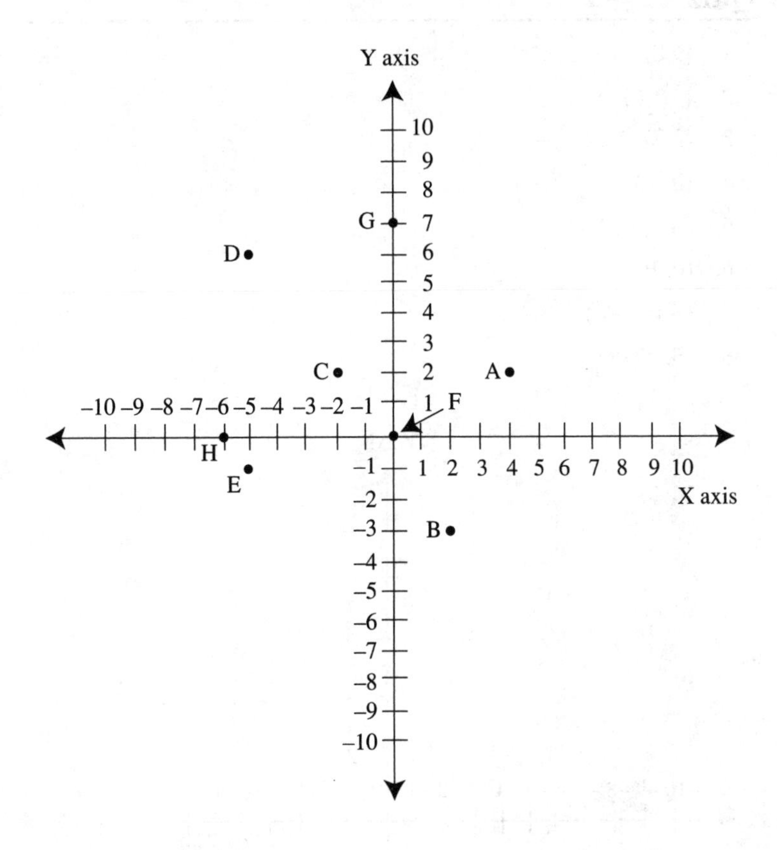

Now try plotting your own points on the graph that has been so thoughtfully provided.

Quiz #2

1. (9,0)
2. (−5,6)
3. (1,2)
4. (0,−4)
5. (−2,−6)
6. (0,0)
7. (−10,−2)
8. (5,−7)

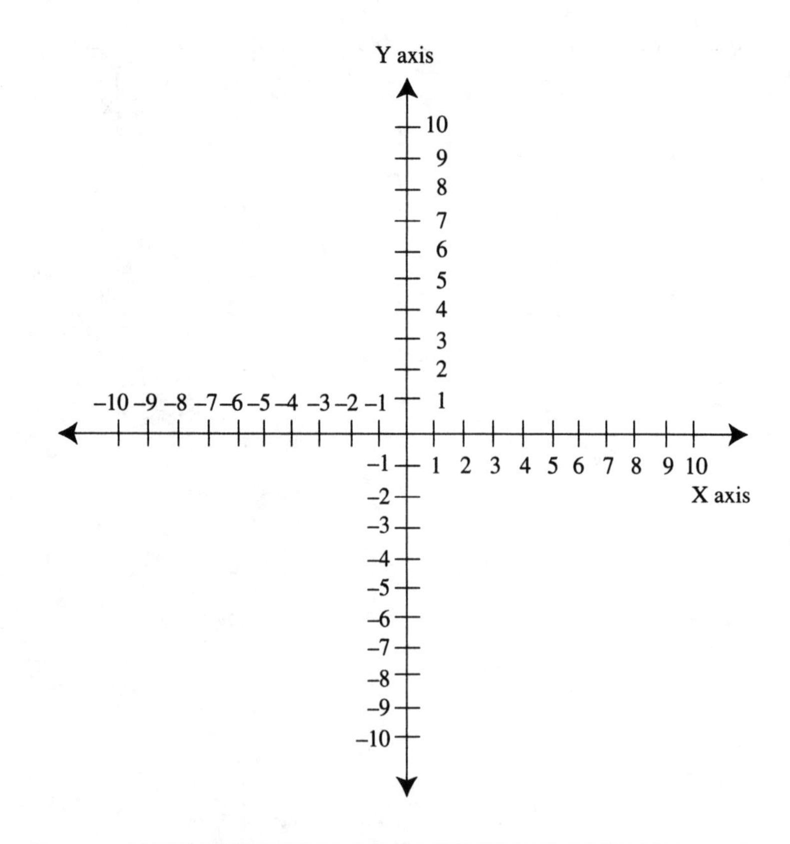

LINES

Now that you've mastered points, let's move on to **lines**. Lines are drawn to connect points. Lines are generally assumed to have no thickness, but they have infinite length. Any two points constitute a line, because you can draw a line through them. Most of the "lines" you see in geometry are actually line segments, or cut up pieces of lines, so they will fit on a page. The lines below are those that illustrate one consistent x or y coordinate.

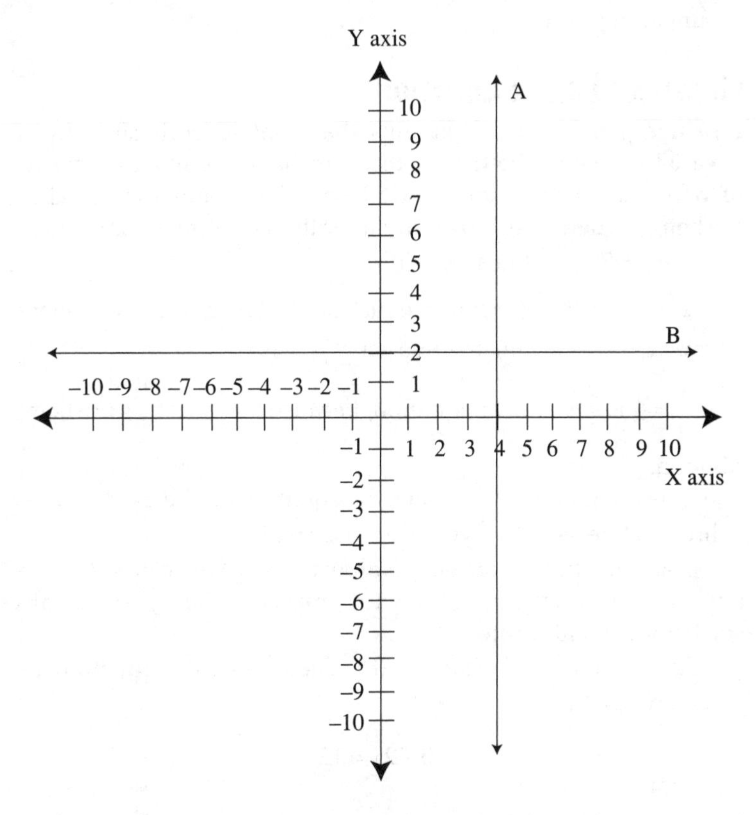

Line A is made up of an infinite number of points, and every point has an x coordinate of 4. Since the x, or horizontal, coordinate is fixed, this line can only run vertically.

Line B is made up of an infinite number of points, and every one of these points has a y coordinate of 2. Since the y, or vertical, coordinate is fixed, this line can only run horizontally.

When one point is fixed, either x or y, then the rest of the infinite points that make up the line are variations on the other co-ordinate, y or x, respectively.

These lines are what you would draw if you were called upon to graph a line for the equation $x = 4$, or $y = 2$. These two lines intersect at (4,2).

You may be thrilled to see that some lines give you a way to integrate your knowledge of algebra, because lines express algebraic relationships. These particular lines are expressions of **linear equations**.

Graphing Linear Equations

Linear equations are equations that contain both an x and a y variable, where both variables are being raised to a power of one (that means not squared or cubed), and the variables are being added or subtracted, rather than multiplied or divided.

$x + y = 5$ is a linear equation.

$x^2 + y = 5$ is *not* a linear equation, because of the exponent.

$y = 3 - x$ is a linear equation.

$\frac{x}{y} = 4$ is *not* a linear equation, because the variables are being divided.

Linear equation also means the equation can be used to draw a line. Here is what you do: $x + 2y = 12$

You first put in an easy value for x. You can substitute any variable you want, but so your graph isn't huge, you usually start with small values.

For instance, say that $x = 0$. Then solve the equation for y, when $x = 0$.

$$0 + 2y = 12$$

$$2y = 12 \quad y = 6$$

You have one point on your line, when $x = 0$, $y = 6$, or (0,6). Plot your point.

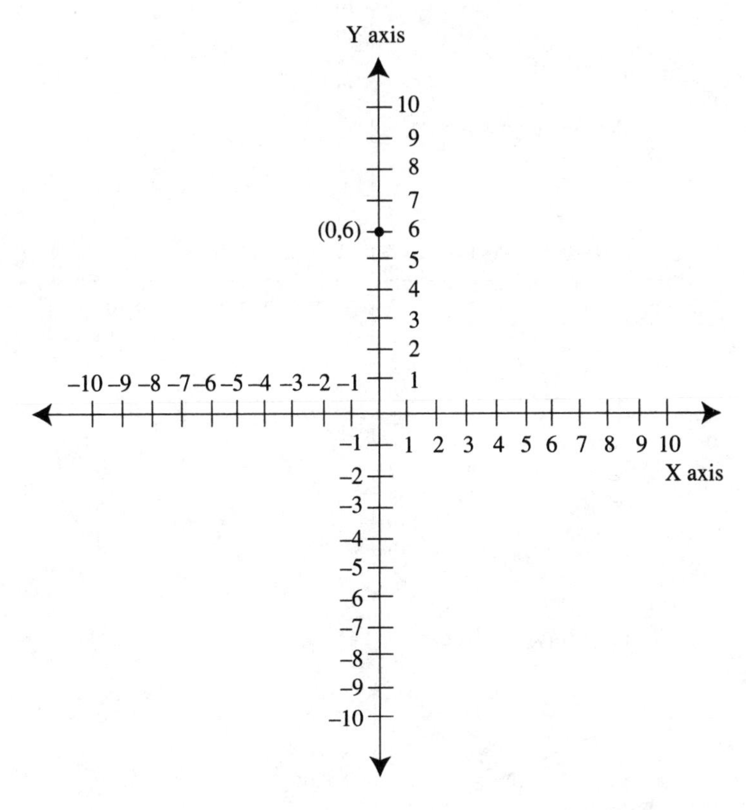

Then, substitute another easy value for x. How about 2?

$$2 + 2y = 12$$
$$2y = 10$$
$$y = 5$$

You have another point for your line, when $x = 2$, $y = 5$, or (2,5). Plot it on that same graph.

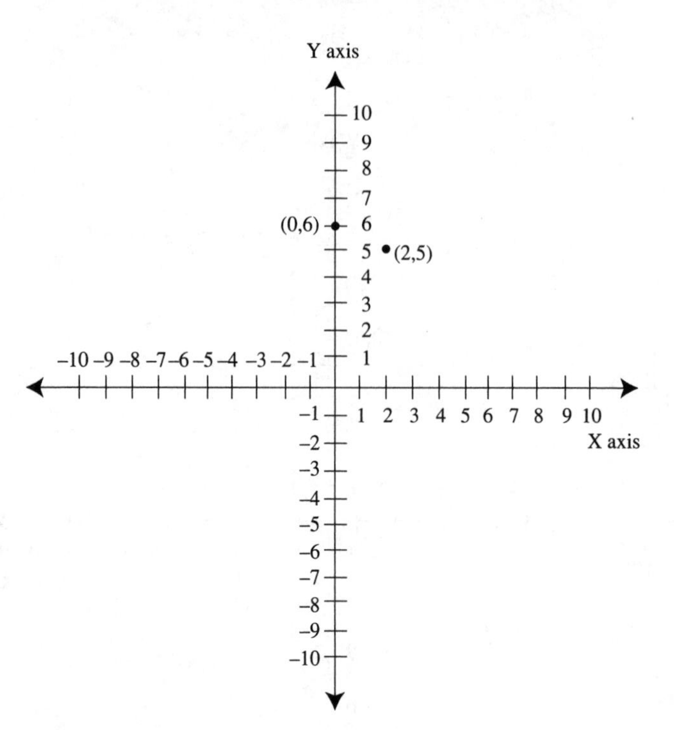

You now have a line.

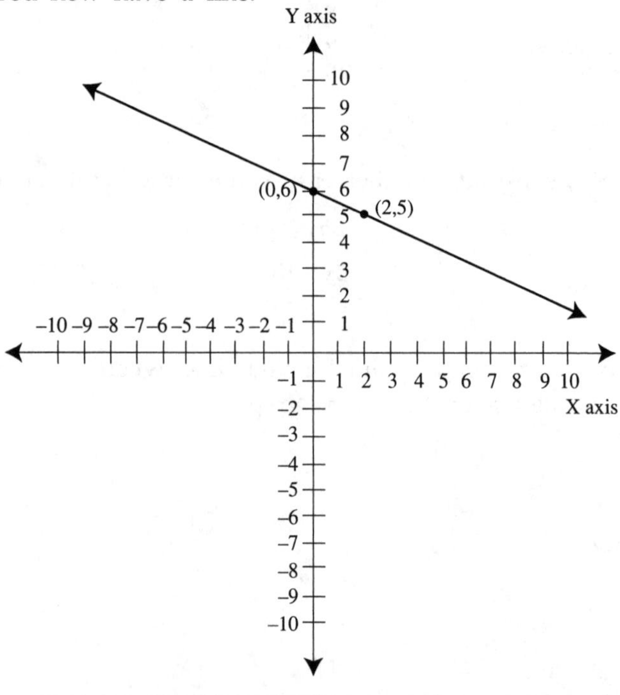

Any two points constitute a line, and you just join them up and draw it right on through. However, in most circles, it is customary to try out a third value for x to check your work.

$$x = 4$$
$$4 + 2y = 12$$
$$2y = 8$$
$$y = 4$$

So your point is (4,4). Does it fall on your line? Well then, you're in business.

A lot of people like to simplify the equation first, to put y in terms of x so it will be easier to find values by substituting. You can do it either way, simplify before or after. Try it both ways and see which feels comfortable.

SLOPE

The **slope** of a line is a way of measuring the steepness of its slant. You can look at it as though a line is a hill; the slope will tell you just how steep the hill is.

What about when it is not a hill? What about when it is totally flat, or totally vertical? Well then you're right; it isn't a hill. If a line is straight vertical or horizontal it has no slope.

The slope of a line is the ratio of **rise** over **run**.

$$\text{slope} = \frac{\text{rise}}{\text{run}}$$

Here is a graph of the line for the equation $y+4=x$.

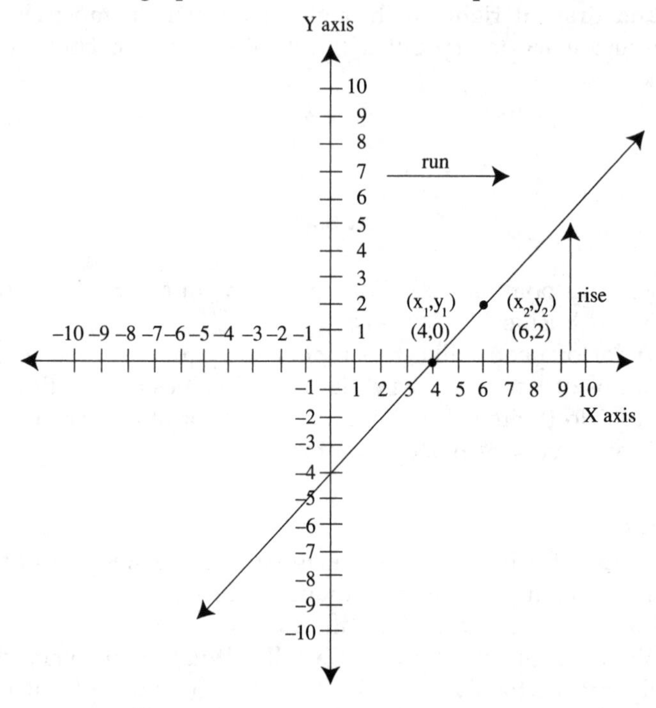

The rise of a line is how much it rises, how much it goes up. The run of a line is how much it goes out. Run is figured by using the X axis. So if you are talking about rise, or how much a line goes up, which axis do you think you will be using? The Y axis, the axis that goes up. The rise is the difference in *y* from one point to the next, also known as the change in *y*. How do you figure out the change in *y*? To figure slope, you are going to find two points on whatever line you have. Like the graph here, you will call them x_1, y_1 and x_2, y_2. To figure the change in *y*, or the rise, subtract y_1 from y_2. To figure the run, subtract x_1 from x_2.

$$\frac{y_2 - y_1}{x_2 - x_1} = \frac{2-0}{6-4} = \frac{2}{2} = 1$$

The slope of the line $y+4=x$ is 1. The slope of a line is referred to as *m*.

Sometimes you will be given an equation and asked to find the slope of its line, and sometimes you will be given two points and asked for the slope.

Try graphing these next equations, and then figure out the slope of each of the lines.

Quiz #3

1. a. $3 + x = y$
   b. What is its slope?

2. a. $2y = x + 1$
   b. What is its slope?

3. a. $x + 3y = 6$
   b. What is its slope?

4. a. $x - y = 2$
   b. What is its slope?

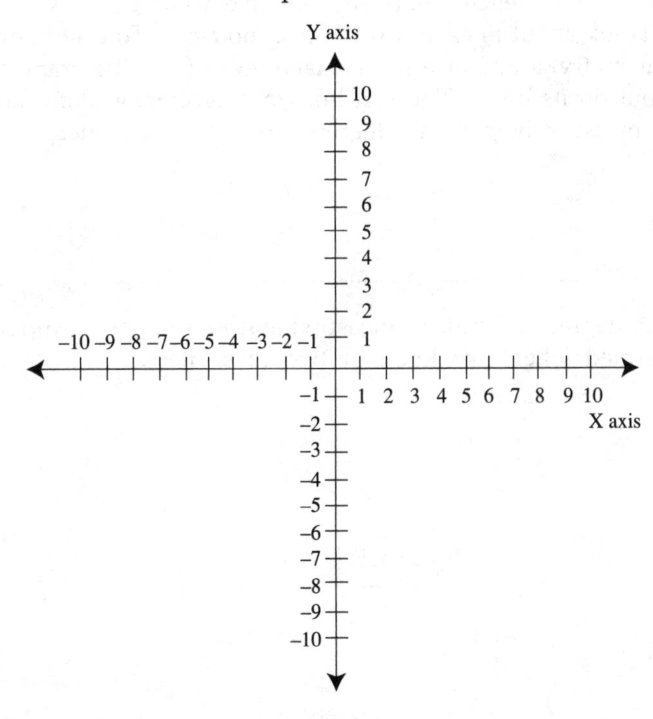

RAYS

A **ray** is a point that extends into a line. Since lines go on forever, so do rays, except that rays only go on forever in one direction.

ANGLES

A protractor is a tool used to measure angles. It looks like this:

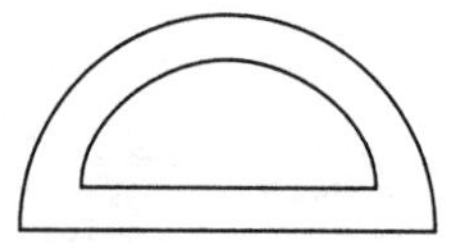

Now you should go out and get yourself a protractor, a really nice protractor, the kind with a french curve and little holes to stencil circles in. Not that we will be using the french curve or the stencils, but why not live well?

Think about lines and rays for a moment. To build angles in geometry, a line or a ray is taken away from the graph and set out on its own. You can line your protractor alongside it and measure how many degrees there are in a line.

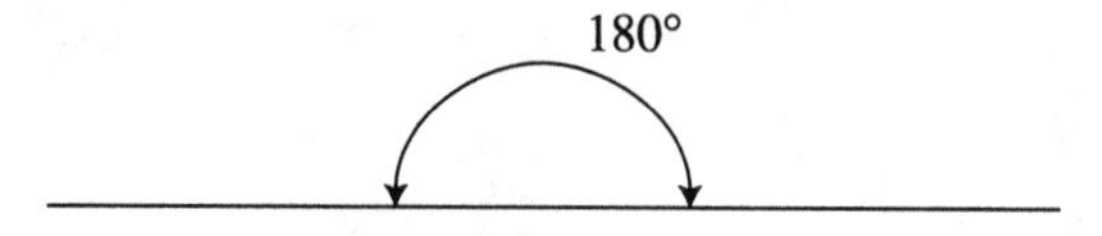

A degree is a unit of measurement in an angle. An angle is formed when two lines, or two rays, meet.

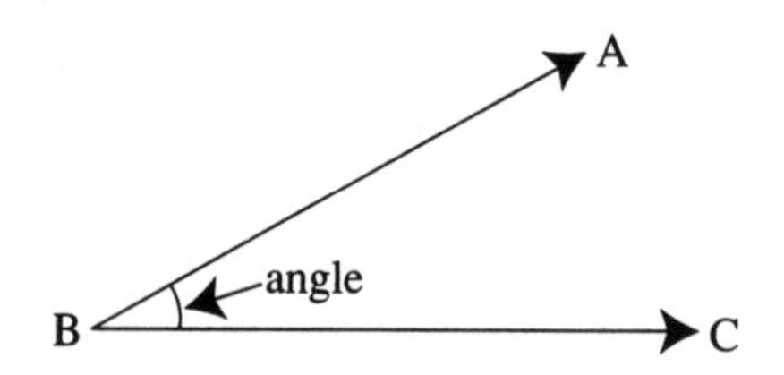

$\angle ABC$ is how the angle on the previous page gets written. $\angle$ is the symbol for angle. The point B is the **vertex** of the angle, the place where the two lines meet to form an angle. When you name an angle, you use the point of the vertex as the middle letter. Sometimes you will also see an angle called by just one letter.

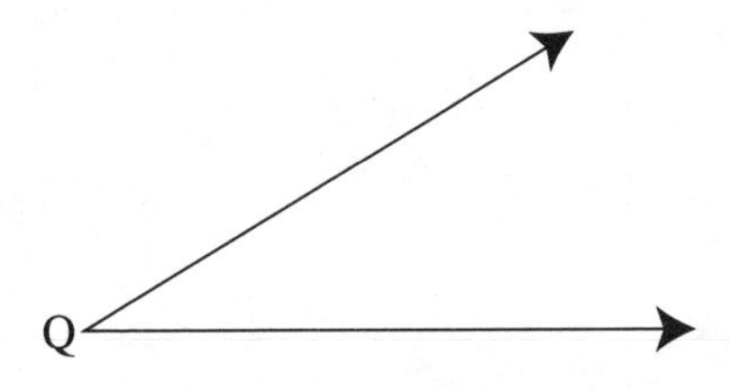

That angle is $\angle Q$.

To measure an angle, line up one of its lines along the flat edge of your protractor, then extend the other line with a ruler if it isn't long enough, and see where it ends on the protractor.

Angle Q measures $30°$.

Angles cannot have negative measures any more than people can have negative heights.

- An angle that is less than $90°$ is called an **acute angle**, like $\angle Q$.
- An angle that is more than $90°$, and less than $180°$, is called an **obtuse angle**, like $\angle D$.

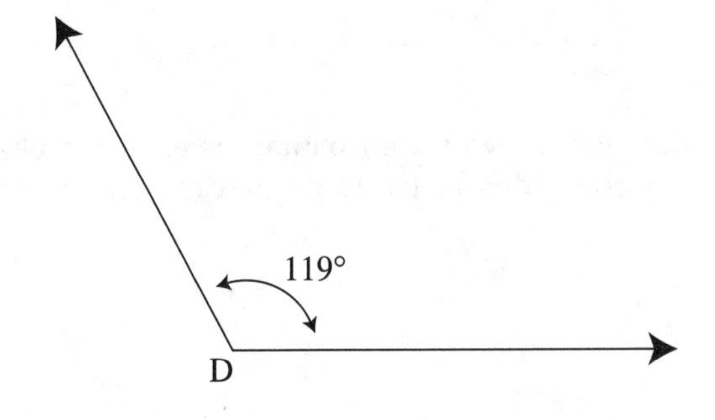

An angle that is exactly 90° is called a **right angle**, like $\angle RST$. You can identify right angles by the little box placed in the vertex of the angle. If you see an angle with that little box, you know immediately that it is a right angle.

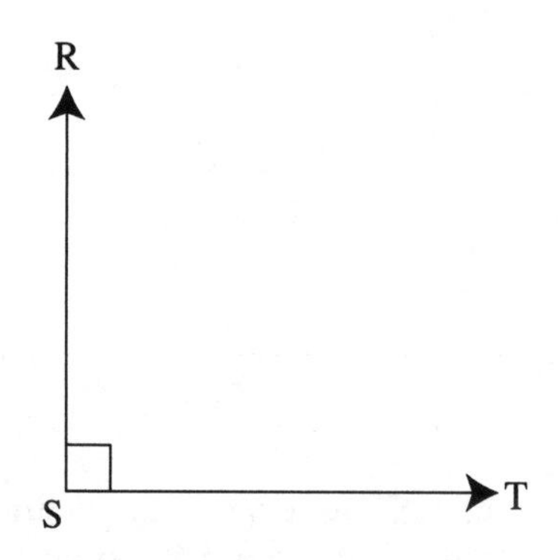

When an angle is bisected, that means it is split into two exactly equal parts.

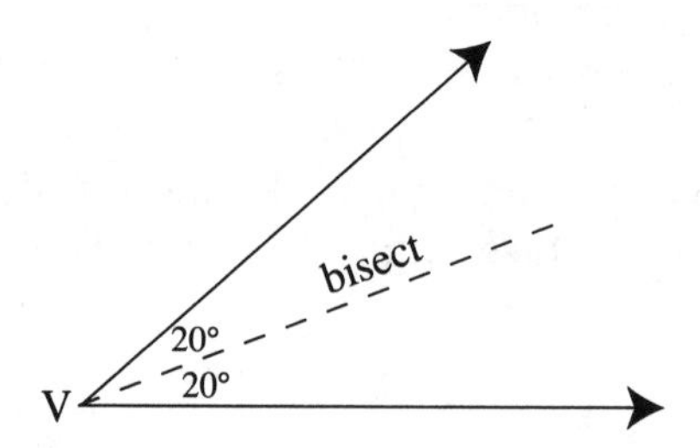

Of course, just as you approximate everything else in the world, it is a good idea to try to approximate angle measurements.

APPROXIMATE THIS:
What is the measure of this angle?

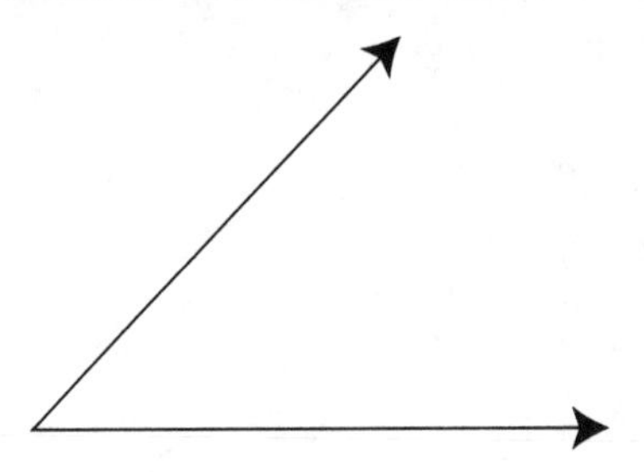

When two lines intersect, they form four angles.

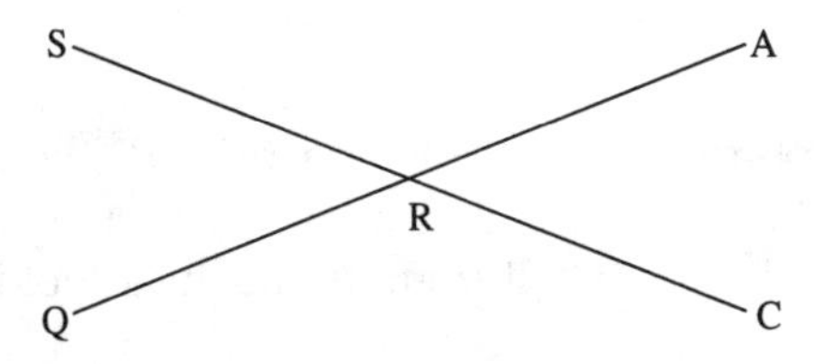

When they do, they form **vertical angles**. Vertical angles are the angles across from one another formed by intersecting lines. In the diagram, $\angle SRQ$ and $\angle ARC$ are vertical. So are angles *QRC* and *ARS*. But look even closer. Put $\angle SRQ$ and $\angle QRC$ together—what do you notice? They form a straight line. That means $\angle SRQ$ plus $\angle QRC$ equal 180° because a line has exactly 180°. Now look at $\angle QRC$ and $\angle ARC$. They form a straight line, too, don't they? That means they also add up to equal 180°. So look at your vertical angles, $\angle SRQ$ and $\angle ARC$. For each of them, if you add $\angle QRC$, they equal exactly 180°.

Think of it as algebra, you can set up an equation.

$$\angle SRQ + \angle QRC = \angle ARC + \angle QRC$$

Subtract $\angle QRC$ from both sides.

$$\angle SRQ = \angle ARC$$

Vertical angles are always equal.

Look at that picture again. You know how $\angle SRQ$ and $\angle QRC$ add up to 180°? That means they are **supplementary**. The term for two angles that add up to 180° is supplementary.

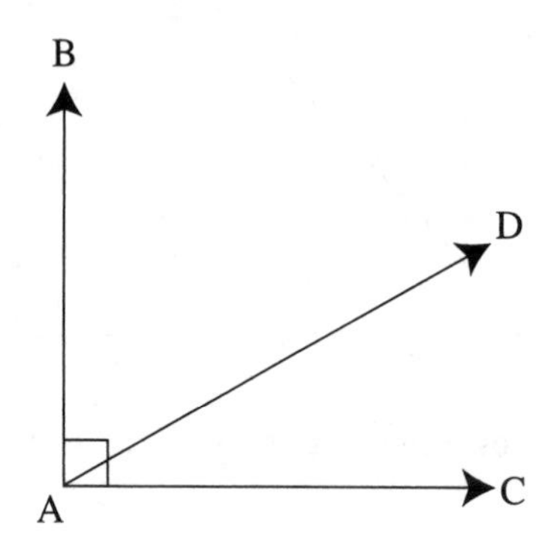

When two angles add up to exactly 90°, they are **complementary**. $\angle BAD$ and $\angle DAC$ are complementary. No, it's not the same as when you tell your friend he's looking fabulous.

Lines That Meet Head On

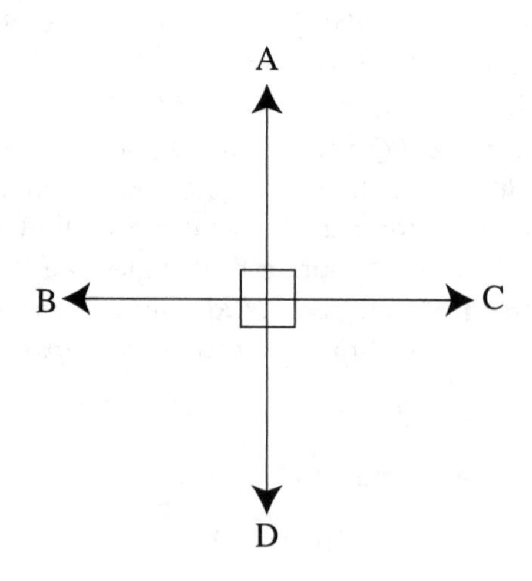

When two lines intersect and form a right angle, the two lines or line segments are called **perpendicular** to one another. The symbol for perpendicular is this: $\perp$. You use it like this: $AD \perp BC$. You read that like this: line *AD* is perpendicular to line *BC*.

LINES THAT DO NOT MEET

When lines run on infinitely, never meet, and are always the same distance apart, they are called **parallel lines**. The sign for parallel is this:

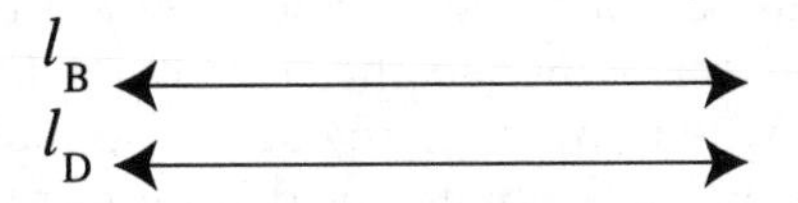

Lines *B* and *D* are parallel. If they ever met, meaning if they ever even started to slant toward one another, they would not be parallel.

When parallel lines are intersected by other lines, angles are formed.

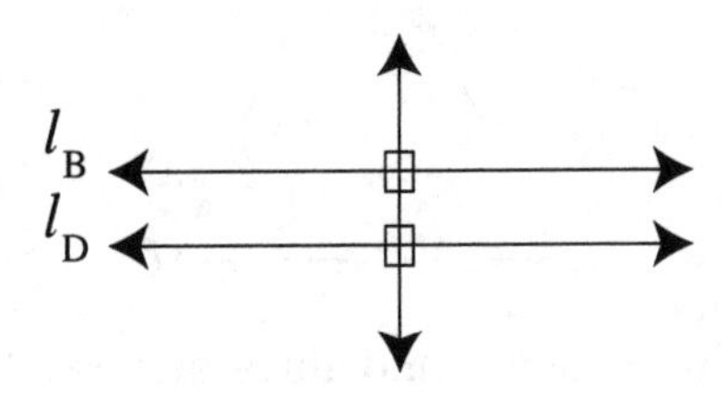

If a third line intersects two parallel lines perpendicularly, what results? Two sets of four right angles.

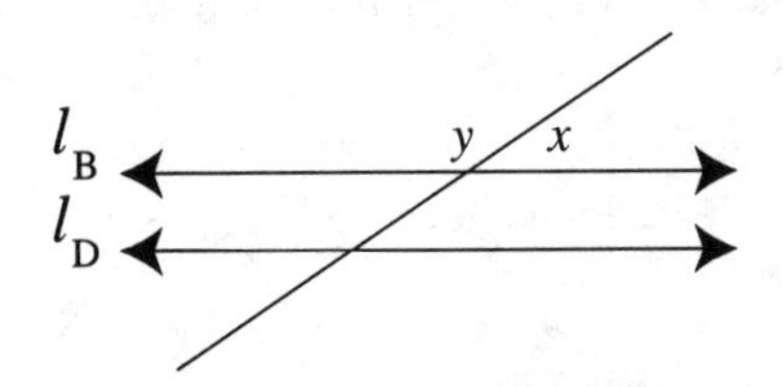

If a third line intersects the parallel lines and it is not perpendicular, there is an acute angle formed, and an obtuse angle formed. In this case $\angle x$ is acute, and $\angle y$ is obtuse. Do you see all the other angles here that are not labeled? They also fall into one of two camps, acute or obtuse. Well here's the thing: all the obtuse angles have the same measure, and all the acute angles have the same measure. Think about your vertical angle information and it will make a lot of sense.

SHAPES

TRIANGLES

You have all these open angles, and they are floating around unattached and annoying, so the next thing to do is look at what happens when you close the angles to make shapes. The first shape we will talk about—read about—is the **triangle**. A triangle is a three-sided figure. Its sign—this is getting to be like the zodiac—is Δ.

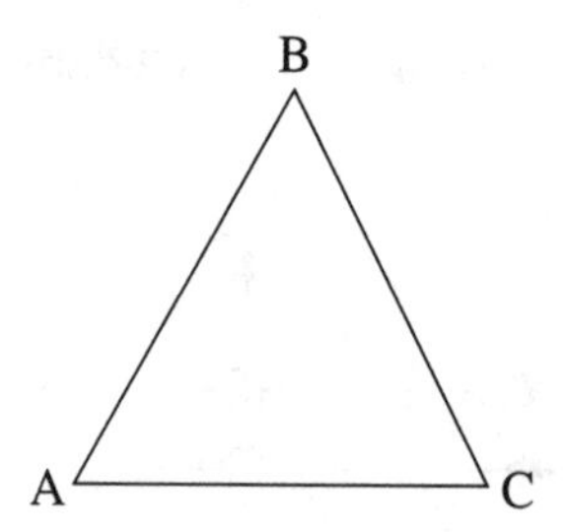

ΔABC has three sides, and three angles. In any triangle in the world, including this one, the angles add up to 180°. Conveniently, you only have to remember one number for lines and triangles. When a line extends outside the triangle, or outside any shape at all, to form an angle, that angle is called an **exterior angle**.

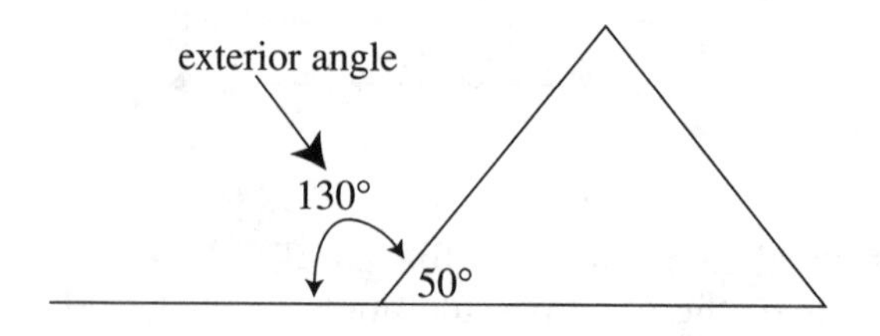

There are certain kinds of triangles you may want to know.

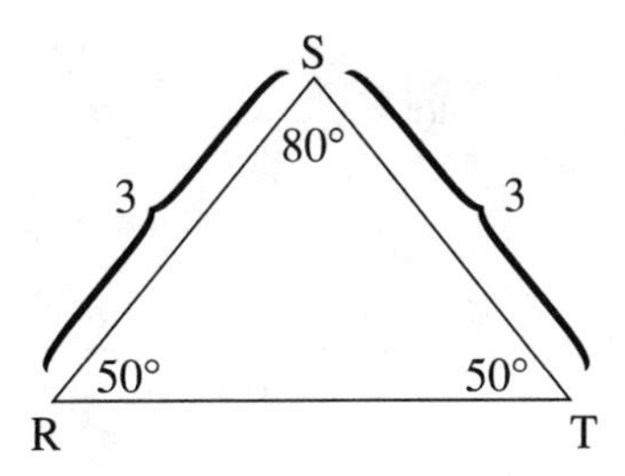

ΔRST is an **isosceles triangle**. That means it has two equal angles, and opposite those two equal angles it has two equal sides.

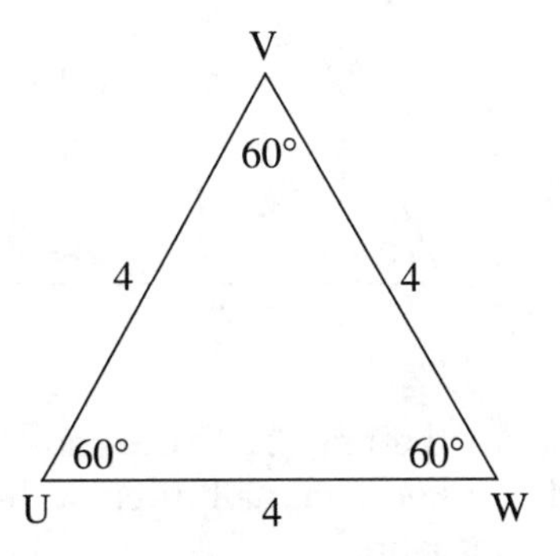

ΔUVW is an **equilateral triangle**. Notice the "equi" prefix? Well it makes sense then, doesn't it, that all the angles and all the sides in this triangle are equal. You may not know what the sides are, they could be 40 feet each.

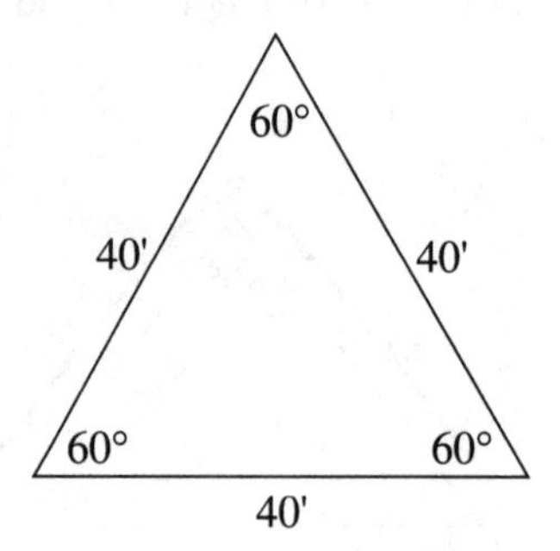

Or four inches each.

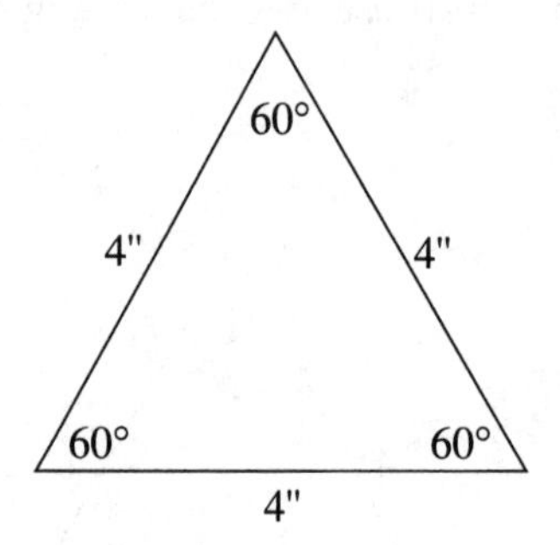

But what about the angles? You know they have to add up to 180°, and you know that all three of the angles are equal, so what is their measure?

$$3\angle = 180^\circ$$
$$\angle = 60^\circ$$

Each angle in an equilateral triangle is 60°.

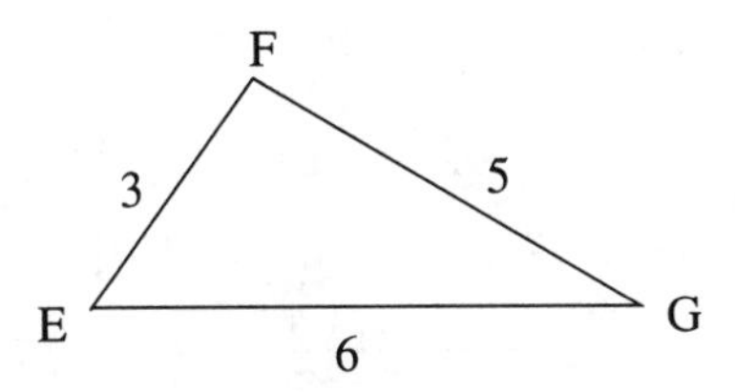

If a triangle has no equal parts, then it is called a **scalene** triangle. ΔEFG is a scalene triangle.

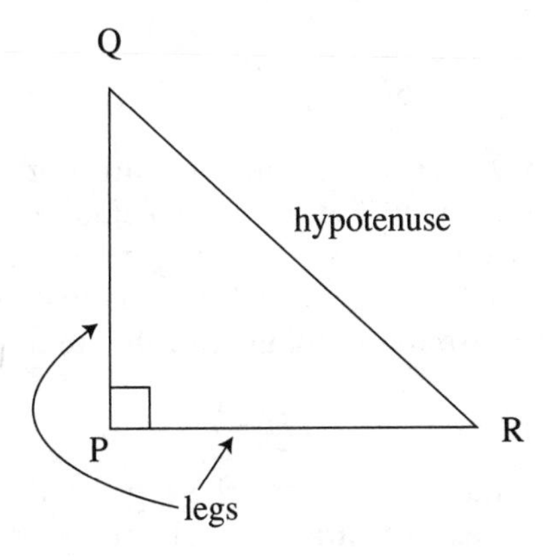

When a triangle has a right angle it is called a **right triangle**. Seems logical, yes? The two sides of the triangle that form the sides of the right angle are called the **legs** of the triangle, and the side of the triangle that is opposite the right angle is called the **hypotenuse**. Right triangles can be recognized by the little square nestling in the vertex of one of the angles, which denotes that the angle is a right angle. When you have two sides of a right triangle, you can always figure out the third side. Why do you care? Well say you have to paint a colorful edging on the top of a wall, flowers or something like that. You need to paint 8 feet up. But built into the wall is this shelflike projection that goes out from the wall about 6 feet.

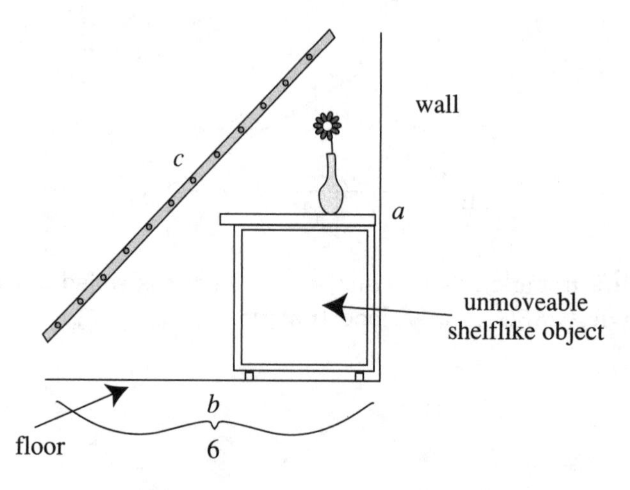

You have a 9-foot ladder but you are not sure it will be long enough, and it's all the way downstairs in the basement, and it's really heavy, so it would be good if you could figure if it is long enough before you bring it upstairs to do your painting. So you use the **Pythagorean theorem**. Which is this:

$$a^2 + b^2 = c^2$$

In this formula, *a* and *b* are the legs of the triangle, and *c* is the hypotenuse. And this formula holds true for all right triangles. Substitute in your numbers for the legs of your triangle, and the ladder will form your hypotenuse.

$$8^2 + 6^2 = c^2$$

$$64 + 36 = c^2$$

$100 = c^2$. To find *c*, take the square root of 100.

$\sqrt{100} = 10$, so the ladder, or *c*, needs to be 10 feet long. Is your ladder long enough? No way, it's only 9 feet, remember? You can forget about the painted flower border on the wall unless you want to go out and buy a new ladder. Anyway, are little flowers on the wall really worth it?

Another way triangles get labeled is when they are **congruent**. Congruent triangles are triangles that have equal sides and equal angles.

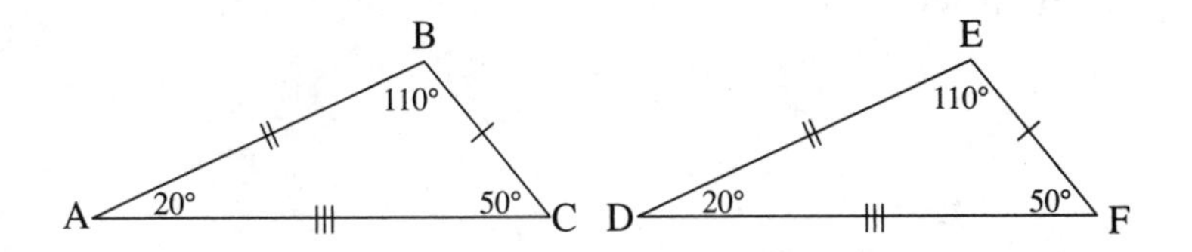

ΔABC and ΔDEF are congruent triangles. The sign for congruent is this: $\cong$.

Triangles are called **similar** when they have equal angles but their sides are not equal; they are just in the same ratio.

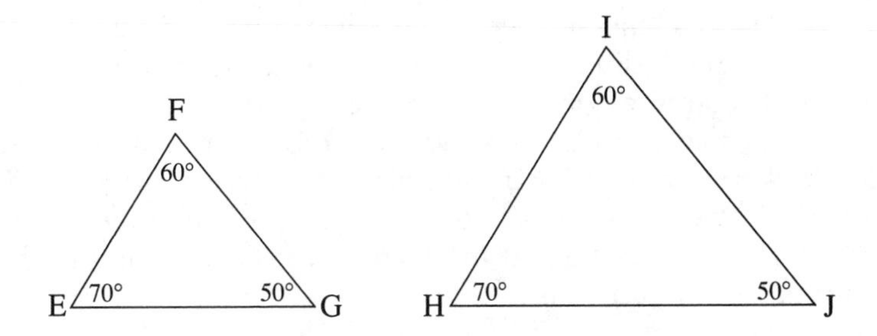

ΔEFG and ΔHIJ are similar triangles. If two triangles are congruent, then they are also similar. But if they are similar, it does not necessarily mean they are congruent, unless the ratio of their sides is 1:1.

What is important about this is that if you know the measurements of one triangle, and you can see that the other triangle is similar, you can find the measure of the other triangle.

For instance, the two triangles here are similar.

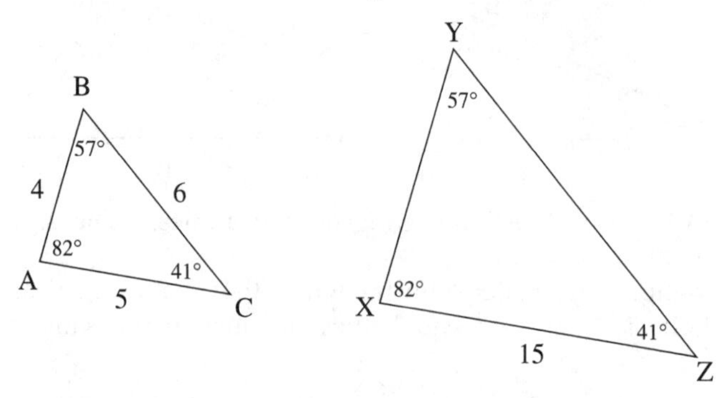

It says that the smaller triangle has sides 4, 5, and 6. If you know the medium side, the side opposite the medium angle, on the bigger triangle is 15, what are its other two sides?

Well you get to use ratios and proportions here, from chapter 3. Check back to page 136 if you feel unsure. 4:5:6 is proportional to ?:15:?? You need to ask yourself what you multiplied 5 by, to get 15. You multiplied by 3, so you need to multiply the other sides by 3 as well.

$$\begin{array}{r} 4 : 5 : 6 \\ \times\ 3 : 3 : 3 \\ \hline 12 : 15 : 18 \end{array}$$

The other sides of the bigger triangle are 12 and 18.

Once you know the sides of a triangle, you can find the **perimeter**. The perimeter of any shape is the measurement of the shape's outside border. It's how long a piece of string you would need to wrap around the edge of a shape.

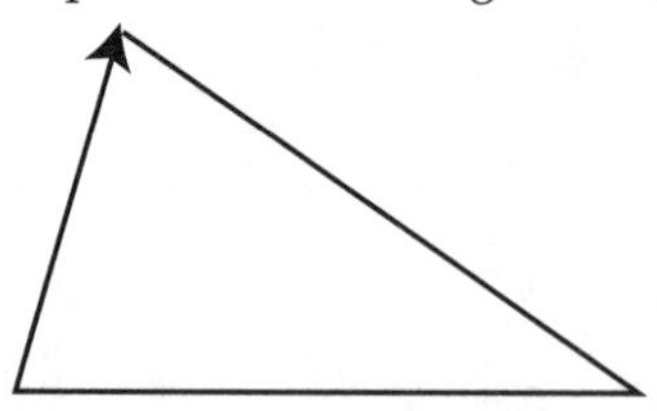

To find the perimeter you simply add the sides together. The perimeter of that larger similar triangle, ΔXYZ is 45. The perimeter of the smaller triangle, ΔABC is 15. Do you notice that perimeters of the triangle maintain the same ratio as the sides of these two triangles? Connections are inspiring, aren't they.

The **area** of a triangle measures the amount of space it covers within its sides.

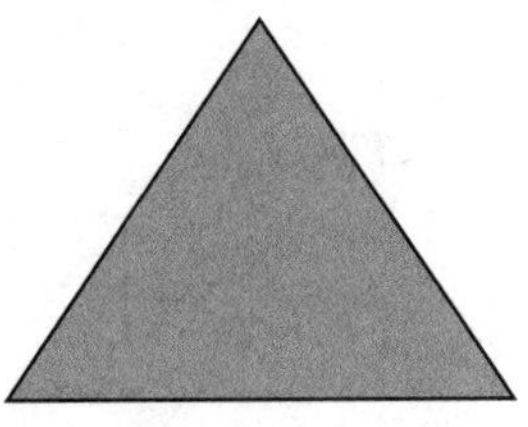

Area is represented by **square units**. A square unit is a square that has sides equal to the unit of measurement the triangle is measured in—inches, feet, or miles.

The area of a triangle is $\frac{1}{2}$ of the base times the height. To find the area of a triangle, you need to know the **base** and the **height**. The height of a triangle is the distance from its highest point to its base, measured by a perpendicular line. The base is the side of the triangle that the height is drawn perpendicular to.

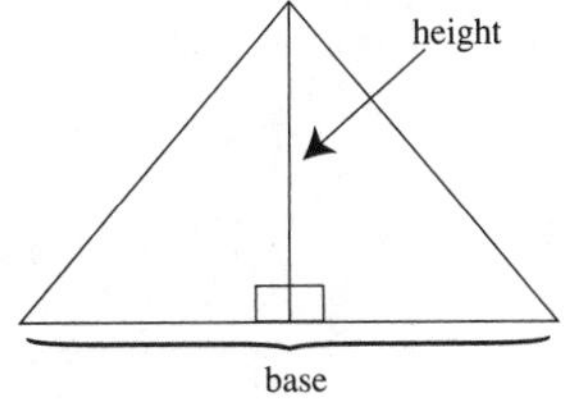

If the base doesn't extend out to the highest point, you can draw a line to extend it.

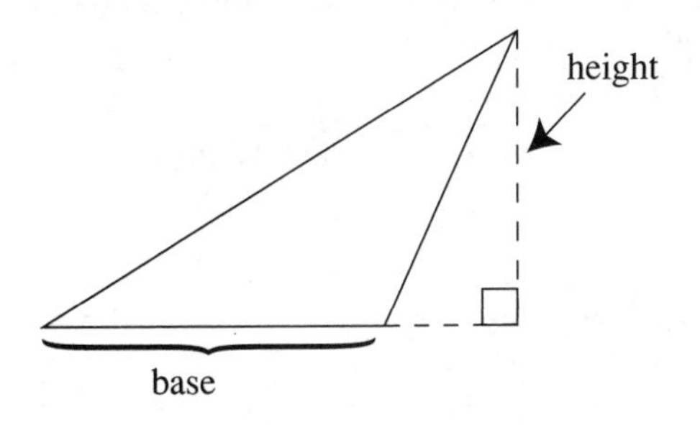

You don't have to draw the height only from one side, you can look at the triangle upside down and still find a height and a base.

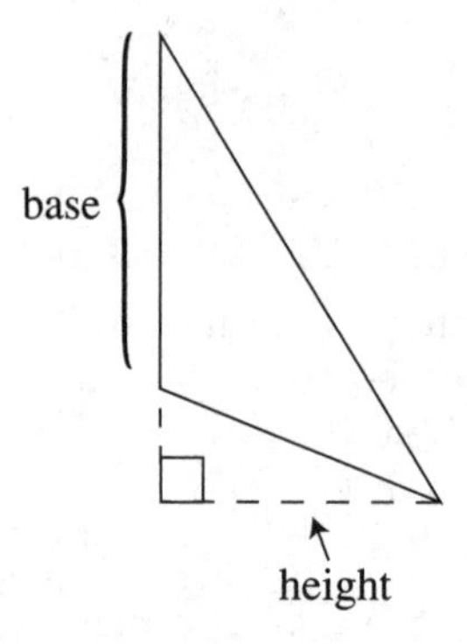

See how they are the same triangle? You use whichever height is easier to find. On a right triangle, the sides meeting at the right angle vertex are the base and the height.

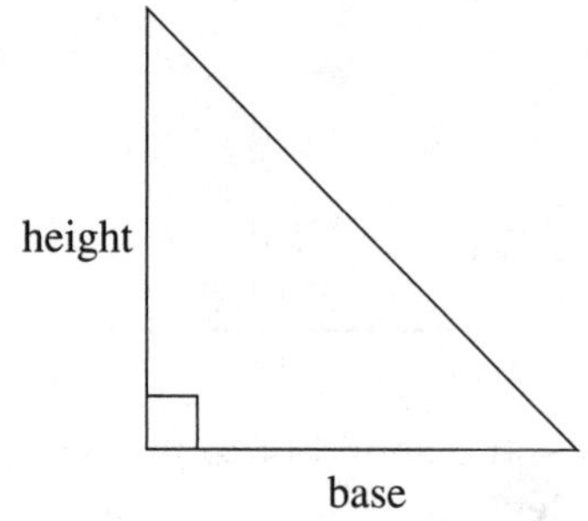

But you can also look at a right triangle as having another base and height.

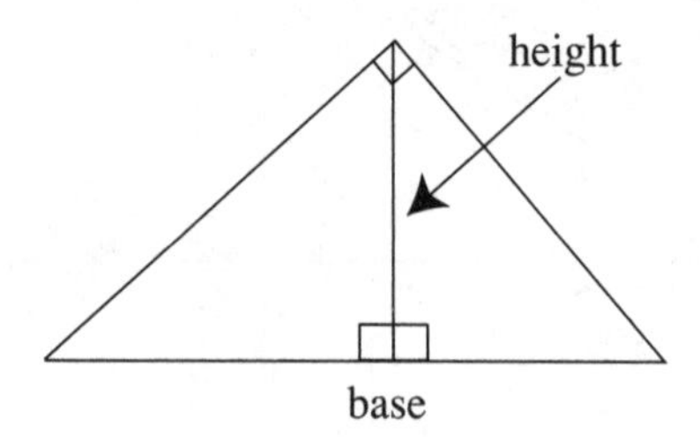

Sometimes finding the measure of the height can be tough. If you're stuck, remember, the height line sometimes forms two smaller triangles. And since the height forms a right angle, these two smaller triangles are right triangles. To find the missing side in a right triangle you can use the Pythagorean theorem.

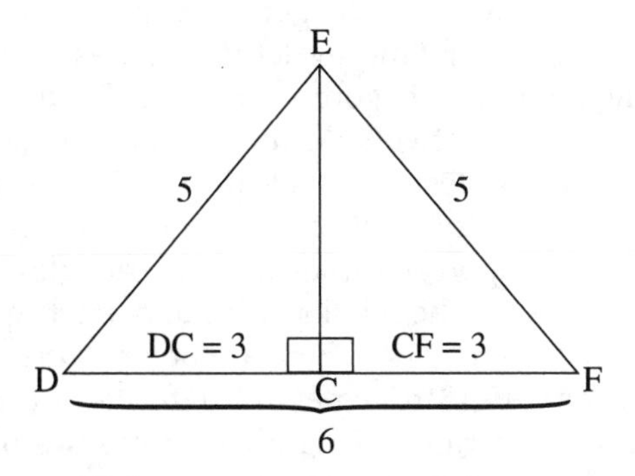

In the triangle above, what is the height? Well, the base is 6, and the sides of the triangle are both 5. The bases of the two smaller triangles are each 3. That means, to find the height, you can set itup like this:

$$3^2 + EC^2 = 5^2$$
$$9 + EC^2 = 25$$
$$EC^2 = 16$$

$EC = 4$. The height of this triangle is 4.

Since area is $\frac{1}{2}$ the base times the height, to find the area of triangle *DEF*, you can take half of the base, or 3, and multiply it by the height, or 4. The area of ΔDEF is 12.

Something to Think About

There are certain triangles called Pythagorean triples. The triangle on the previous page is made of a Pythagorean triple, the 3:4:5 triangle. Any right triangle that has sides in a ratio to this, like 6:8:10 or 9:12:15 is also a Pythagorean triple, and it can save you a lot of time by letting you escape from working out the Pythagorean theorem. Because even if you only have two sides of a right triangle, and they are the two legs and they are 6 and 8, you know the hypotenuse is 10 because it is a version of a 3:4:5 triangle. Pythagorean triples are right triangles whose sides, because of the relationships of perfect squares, page 154, are integers. There are other Pythagorean triples, the 5:12:13 triangle and the 7:24:25 triangle. If you notice a right triangle with the sides or legs in one of these ratios, remembering them can make your life easier. And remember, the biggest side in all of these will always be the hypotenuse. Why? Because sides opposite angles in a triangle are proportional to the angles, and the biggest angle in a right triangle has to be the 90° angle, right? Otherwise the triangle would have more than 180°.

Answer the following triangle and angle questions, and have a good time doing it.

Quiz #4

1. In ΔCDE, what is the measure of angle EDC?

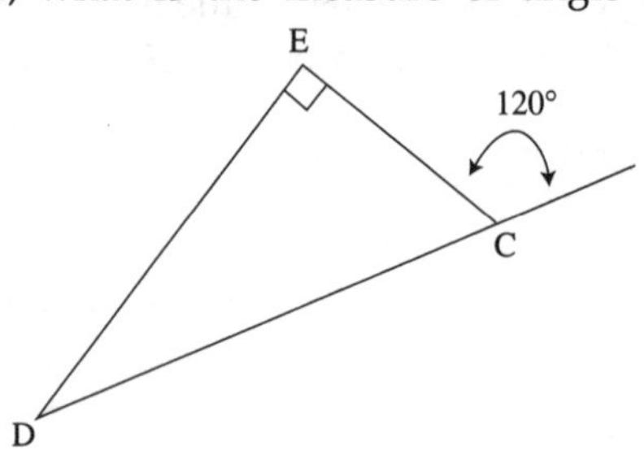

2. In ΔCDE (at the bottom of page 238) what would be the measurement of the two angles formed if angle EDC were bisected?

3. What is the area of ΔEFG?

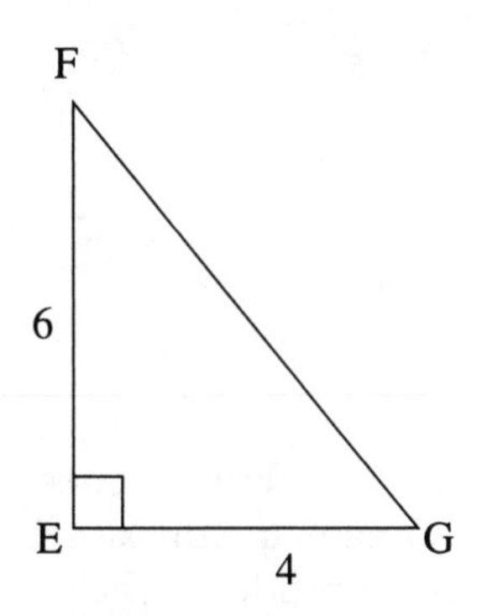

4. What is the measurement of $\angle SRZ$ in ΔQSZ?

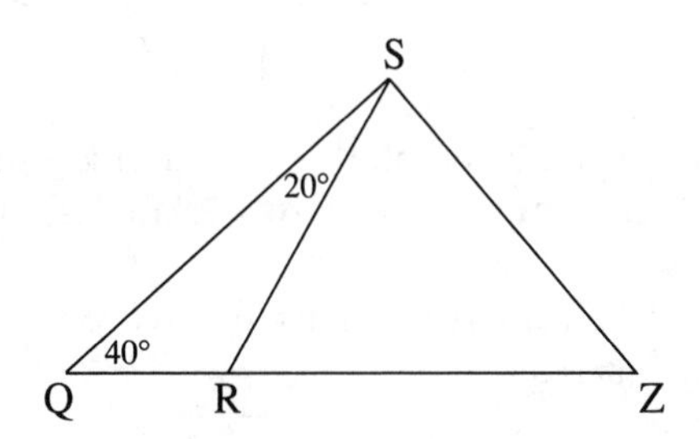

5. What is the perimeter of ΔKLM?

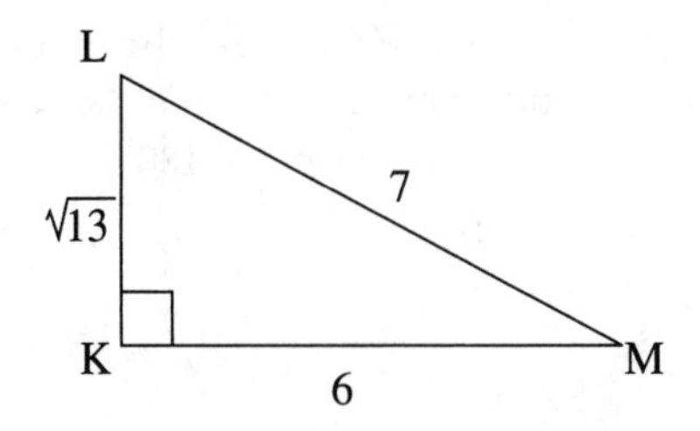

6. What is the area of ΔKLM (at the bottom of page 239)?
7. Are ΔWXZ and ΔDEF similar? Are they congruent?

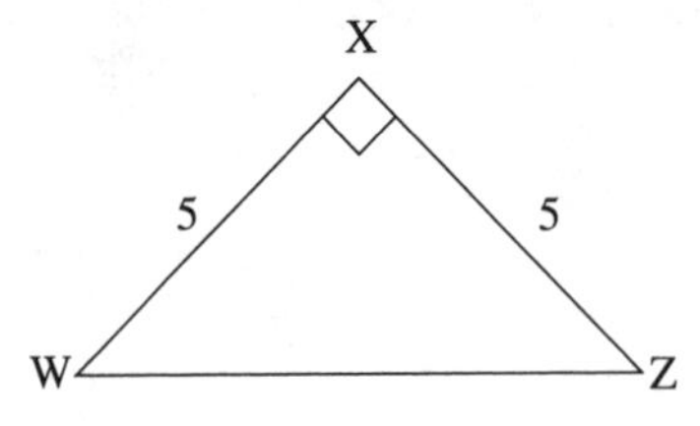

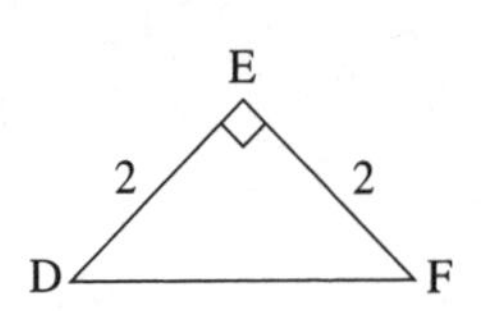

8. What is the area of ΔWXZ?

QUADRILATERALS

Remember quadrants? The four sections of the graph? Well "quad" comes up again here, **quadrilaterals**. Quadrilaterals are enclosed figures with four sides. They can be cut in half to form two triangles.

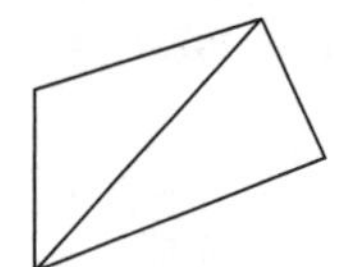
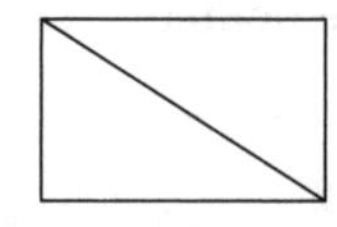
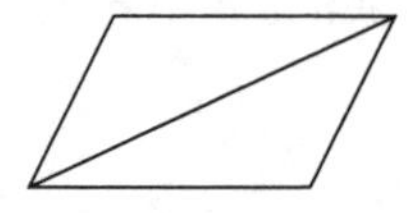

Since you know all the angles of a triangle measure 180°, if a quadrilateral is made up of two triangles, then it has to measure $2 \times 180°$, or $360°$.

A **trapezoid** is a quadrilateral with two parallel sides, and two nonparallel sides.

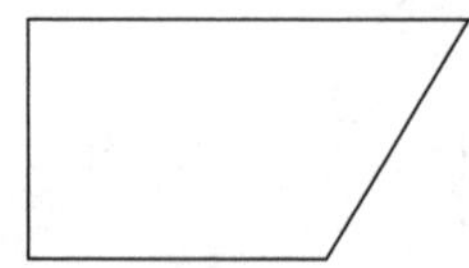

A **parallelogram** is a quadrilateral with two sets of parallel sides. Its diagonally opposite angles, *B*,*D*, and *A*,*C*, are equal. Angles on the same side add up to 180°.

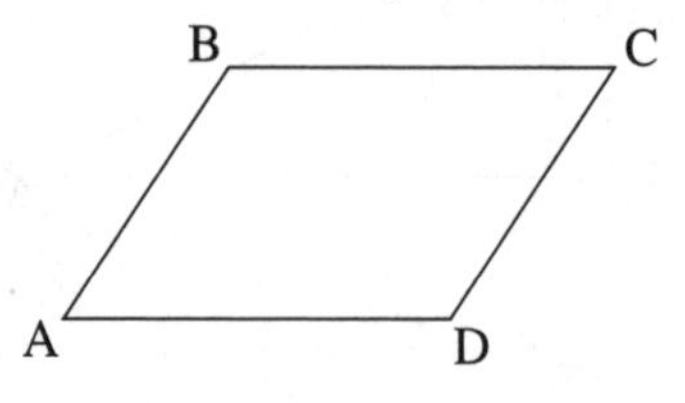

A **rhombus** is a parallelogram with equal sides. It's like a smashed square. It has the same angle properties as any other parallelogram.

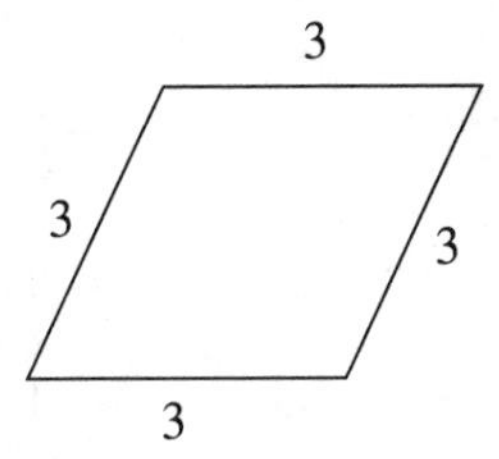

A **rectangle** is a parallelogram with all angles equal to 90°.

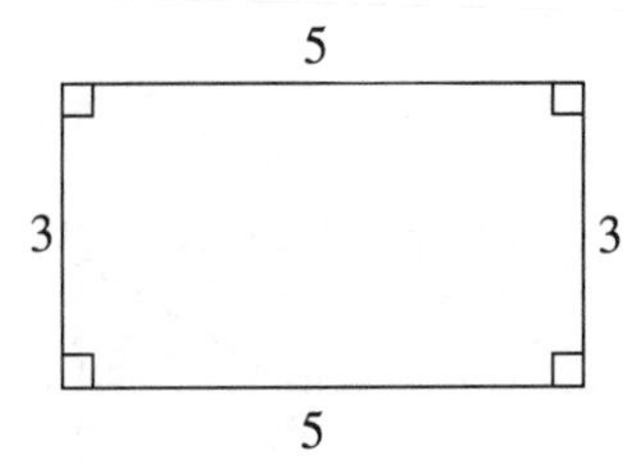

A **square** is a rectangle in which all the sides are equal.

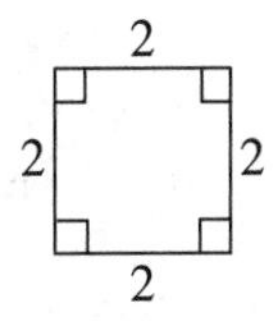

Some of these shapes are subsets of one another. For instance, a square is a type of rectangle, so all squares fit the definition for rectangles. But not all rectangles are squares, because some rectangles may not have all sides equal.

Most of the time, when you are asked questions on standardized tests about four-sided figures, it's in reference to squares and rectangles.

To find the area of a rectangle you multiply base times height. In a rectangle or a square, these are also called **length** and **width**. The length refers to the longer side, and the width refers to the shorter side.

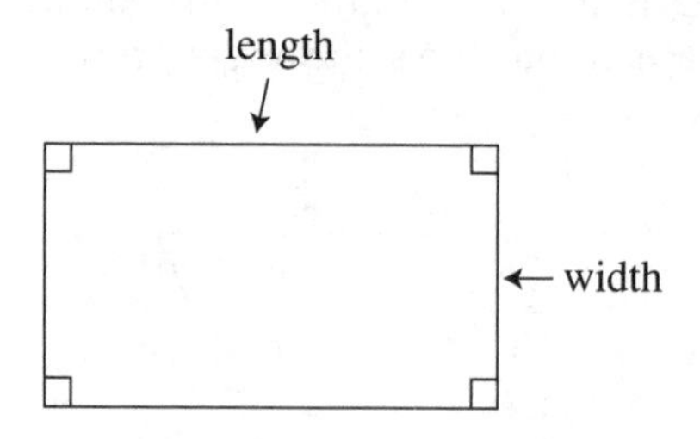

The way to find area makes sense, because think about a rectangle being made up of two triangles for a minute. The lines from opposite vertices of a quadrilateral are called **diagonals**.

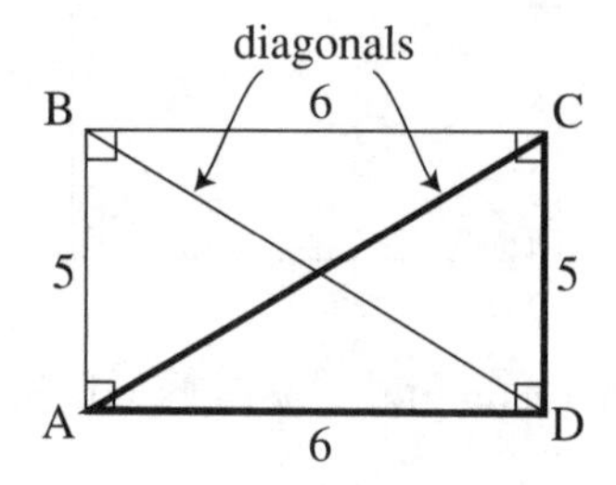

The area of each of the triangles that make up rectangle *ABCD*, ΔACD and ΔABC, is 15. The base of each is 6, the height of each is 5. Area of a triangle is $\frac{1}{2}b \times h$, or $3 \times 5 = 15$.

The area of the rectangle is the area of the two triangles put together, isn't it? So the area of the rectangle is $15 + 15$ or 30. That's the base times the height, except this time you don't divide by 2.

And what is the perimeter of the rectangle? Add the sides. $5 + 6 + 5 + 6 = 22$.

You could look at it as $2(5) + 2(6) = 22$, since you have two pairs of equal sides.

Just as with any rectangle, the area of a square is length times width, or base times height, but since in a square all sides are equal, the formula for area of a square is usually expressed s^2 or side squared.

The area of other quadrilaterals is also figured by multiplying base times height, but since they may not have right angles, you need to find the height of the shape before you calculate.

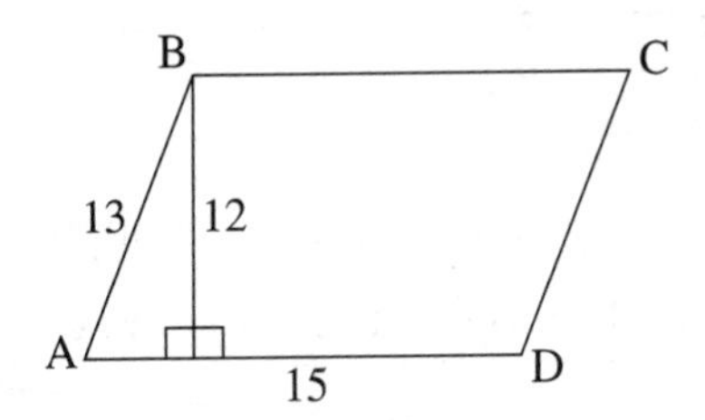

The area of this parallelogram is base times height. The base is 15. What is the height? The side has length 13, but the side of a parallelogram is not equal to the height because the side is not perpendicular to the base. The height is a line that goes straight down and is perpendicular to the base. In this case, the height is 12, and the area is 15 (the base) times 12 (the height).

$15 \times 12 = 180$. You can use distributive property here, can't you? 15 times 2 is 30, and 15 times 10 is 150. $150 + 30 = 180$.

Always practice throwing numbers around in your head trying to multiply them different ways; it makes patterns and connections appear before your eyes. By the way, what is the perimeter of this parallelogram? Add up all the sides, and you have it.

$$15 + 13 + 15 + 13 = 56$$

The perimeter of the parallelogram is 56. But one of the great things about a lot of these four-sided figures is that they have pairs of equal sides. So (2×15) plus (2×13) is another way of looking at the perimeter.

$$30 + 26 = 56$$

And of course, you should approximate first, and see how closely your answer fits with your approximation.

Finding the area of a trapezoid is a little different, because the sides are all different lengths.

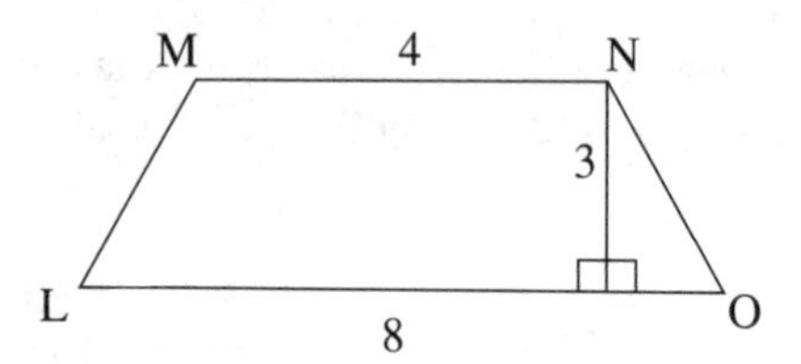

The area of a trapezoid is $\frac{1}{2}$ the sum of the top and bottom, times the height. The top is 4 and the bottom is 8. The sum of these two sides is 12. Half the sum, or 6, times the height, or 3, is 18. The area of trapezoid *LMNO* is 18.

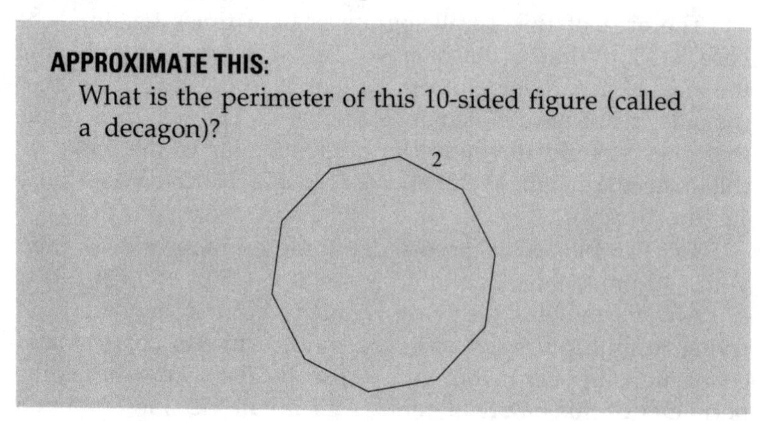

Other Polygons

Shapes other than triangles and quadrilaterals are called **polygons**. "Poly" means many, just like polynomials. They are also called by a numbered type prefix, followed by "gon," for more specificity. For instance, *penta*gons are 5-sided figures, like the United States defense building.

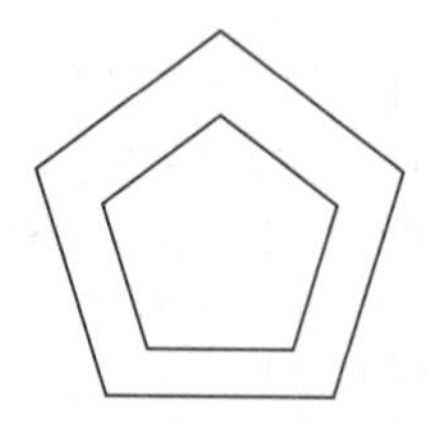

*Octa*gons, or 8-sided figures, are like your friendly local stop sign.

Also, all polygons can be similar, just as triangles are similar. If two polygons have the same number of sides and the same angle measurements, then their sides are proportional and they are similar.

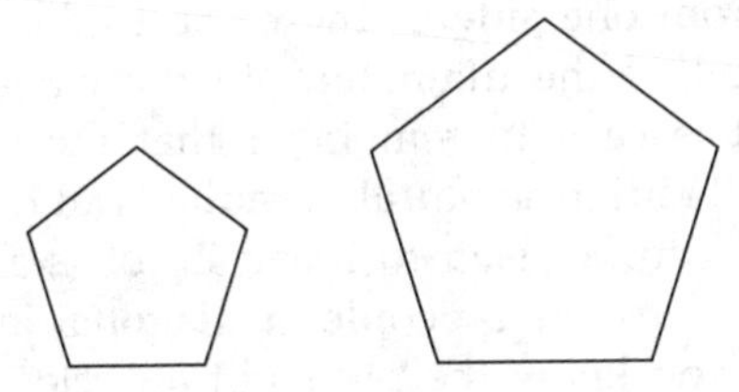

These two pentagons are similar.

CIRCLES

Circles are perfectly round shapes. Yes, we know you knew that but we might as well put all the information on the table. You can use a compass to draw a circle, or you can trace the lid of any old jar. No need to get fancy.

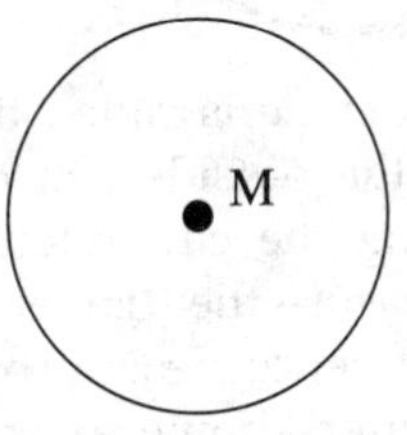

To have something be a circle, or perfectly round, all the points on the edge need to be exactly the same distance away from the **center**. The center of a circle is the point in the middle. This circle has center *M*.

Draw a straight line through the center of a circle and notice that each side of the line, like all lines, equals 180°. Since both sides of the line are included in the circle, a circle measures 360°.

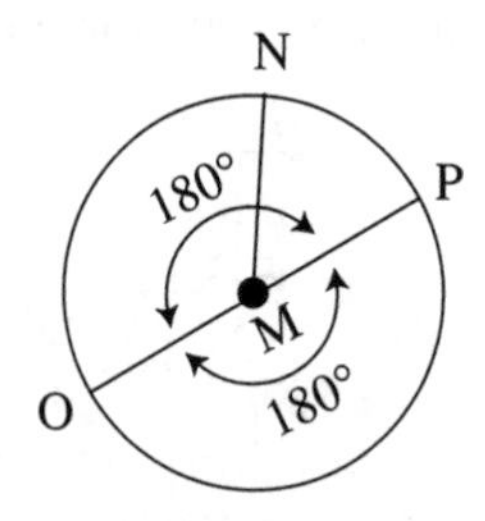

Any line from the center of the circle to any point on the edge of the circle is called the **radius**. In the circle above, *MN* is a radius.

Any line from one side of the circle to the other, through the center, is called the **diameter**. In the circle above, *OP* is a diameter. It is easy to remember that the diameter is the same length as 2 radii (the plural of radius) put together because the prefix "di" can be used to mean 2, for example, a **di**alog is a discussion between 2 people, a **di**agonal joins 2 opposite vertices. So if you know the radius of a circle, you also know that twice that radius is the diameter.

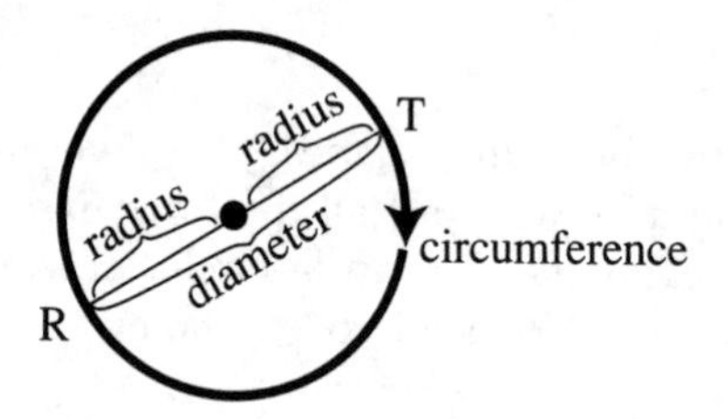

The perimeter of a circle is called the **circumference**. The measurement around the outside rim of a tire is the circumference. Try measuring the circumference of a circle with a tape measure, then measure the diameter with a tape measure. Now divide the circumference measure by the diameter measure. You get 3.1416 . . . better known as π. To calculate with π (spelled "Pi" pronounced like "Pie"), you can use 3.14, even though this isn't the most accurate representation of π in the world.

π is the ratio of the circumference to the diameter. The ratio of diameter to circumference is equal for all circles, so all circles are similar. Since you may not want to measure around circles all the time, you can use π to express measurements.

If you want to know the circumference when you have only the diameter, multiply the diameter times π and you have the circumference. It works the same way as any ratio does. And if you have the circumference and you want the diameter, divide by π and there you have it.

$$\text{Circumference} = \text{Diameter} \times \pi$$

$$\text{Diameter} = \frac{\text{Circumference}}{\pi}$$

Most math books will call this $2\pi r$, rather than πD (for diameter).

To find the area of a circle, you use π again. Multiply π by the square of the radius and you have the area.

$$\text{Area} = \pi \times \text{radius}^2$$

CONSIDER THIS

The circumference of a circle is much bigger than it looks, in fact it is more than three times the size of the diameter, isn't it? If you take a standard kitchen glass and measure it, the circumference will almost always be greater than the height of the glass. You can win lots of bets with people about this because it just doesn't look that way. Remember to have a tape measure or a piece of string to mark to prove your point though.

APPROXIMATE THIS:

How many grooves are there around the circumference of a quarter?

There are some other pieces of circles, and things having to do with circles.

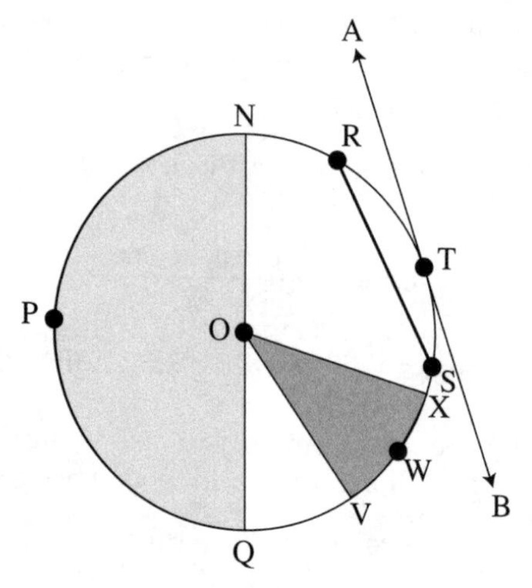

A **chord** is a line segment from any edge of the circle to any other edge. *RS* is a chord.

A **semicircle** is exactly one half of a circle. *NPQ* is a semicircle. The *P* tells you the semicircle in question is on the left side, rather than *NQ* which could be either semicircle.

An **arc** is a section of the circumference of a circle. A **minor arc** is an arc less than 180°. *RTS* is an arc. Arcs are proportional to the angles they correspond to: the bigger the arc—just the cutting a slice of a pie—the bigger the angle at the point of the pie wedge. *VWX* is also an arc.

A **sector** is the area covered from an arc to the center of a circle, bounded by two radii. *VOX* is a sector. Sectors, like arcs, are proportional to the angles they correspond to. And for both sectors and arcs, minor refers to those sectors or arcs corresponding to less than 180°.

A **tangent** is a line outside the circle that meets the circle at only one point. *ATB* is a tangent.

MEASUREMENTS OF PARTS OF CIRCLES

To find the values of arcs and the areas of sectors, you will be helped by everything you have learned so far.

What is the length of arc *ABC*?

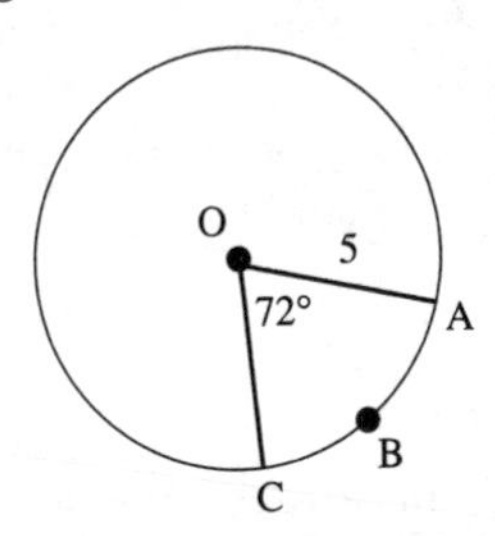

To find the length of arc *ABC*, you need to remember that it's a section of the circumference. The easiest thing to do is find the circumference first. The radius *OA* is 5, and circumference is π times the diameter, so you need the diameter. The diameter is equal to 2 radii, or in this case, 10. And the circumference is π times 10, or 10π. (Math people like to put the number before the π.) To find the section of the circumference that *ABC* is, look at angle *AOC*. This angle, as with any angle that corresponds to an arc or a sector, is proportional to its arc. "But," you say to yourself, "this angle is 72°! That's such a weird number of degrees." See how it corresponds to the number of degrees in the circle.

$$\frac{72}{360} = \frac{36}{180} = \frac{18}{90} = \frac{9}{45} = \frac{1}{5}$$

The angle is $\frac{1}{5}$ of the circle. This means the arc is $\frac{1}{5}$ of the circumference.

$\frac{1}{5} \times 10\pi = 2\pi$. The length of the arc is 2π.

And guess what, the same thing applies to sectors and areas.

What is the area of sector *DEF*?

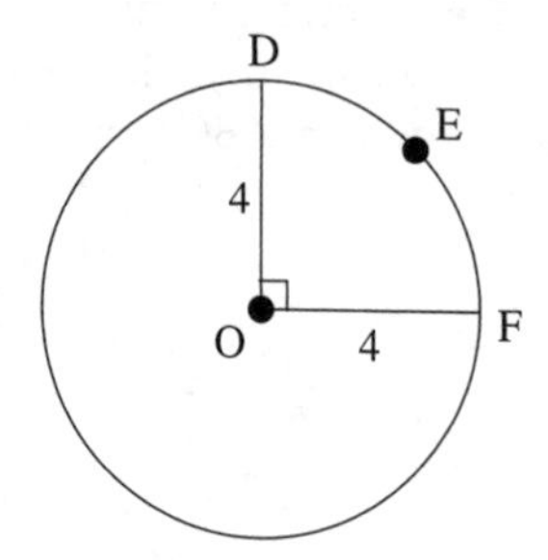

How do you find the area of the sector? Well, the sector is a section of the whole circle, right? To find the area of a section, a fine idea is to find the area of the whole.

$$\text{Area} = \pi \times \text{radius}^2$$
$$\text{Area} = \pi \times 16 \text{ or } 16\pi$$

You know arcs of circles are proportional to the angles they correspond to, so the area of this sector is proportional to the area of the circle. You also know that the angle of this sector is $90°$. The area of the sector is proportional to the area of the whole in the same way that $90°$ is proportional to $360°$.

$$\frac{90}{360} = \frac{9}{36} = \frac{1}{4}$$

The sector's area is $\frac{1}{4}$ of 16π, or $\frac{1}{4}16\pi = 4\pi$.

The area of sector *DEF* is 4π.

This way of looking at sectors can also help you find the length of some chords. In this case, the chord *DF* can be drawn in to form the hypotenuse of the right triangle inside the circle, ΔODF.

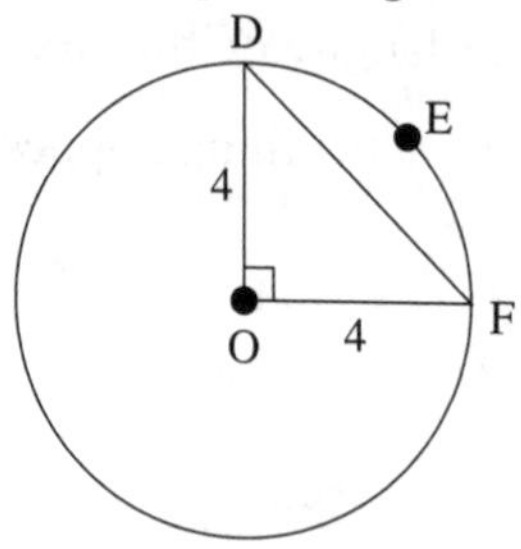

To find chord *DF*, find the length of the hypotenuse by using the Pythagorean theorem.

$$4^2 + 4^2 = DF^2$$
$$32 = DF^2$$

$\sqrt{32} = DF$. You can factor now, if you want. $DF = \sqrt{32} = \sqrt{16} \times 2 = 4\sqrt{2}$.

Shade Problems

This whole sector idea also brings up a lot of other problems that scare people—shade problems. Shade problems are those problems where you are asked to identify the area of a shaded region. These are easy, actually; all you do is find the area of the larger shape, find the area of the smaller shape, and subtract so you find the shaded region. Take a look.

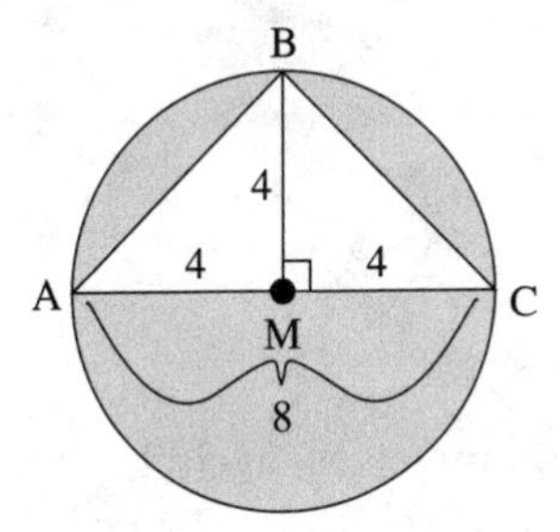

The larger shape here is the circle. The area of a circle is the radius squared times π.

The radius here is 4, so the area of the circle is 16π. To find the area of what is shaded, you have to subtract the area of the triangle from the area of the circle. So what is the area of the triangle?

Well, the base of the triangle also serves as the diameter of the circle, so the base is 8, and the height is the radius of the circle, so the height is 4. Area is $\frac{1}{2}$ the base times the height, or 4×4, or 16.

The shaded region is the circle, 16π, minus the triangle, 16. $16\pi - 16$ is your answer.

Try the following shape questions:

Quiz #5

1. What is the area of ΔDEF?

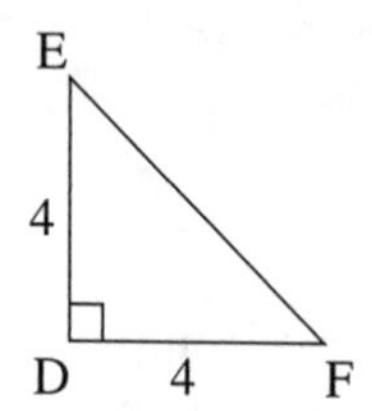

2. What is the area of semicircle *GHI*?

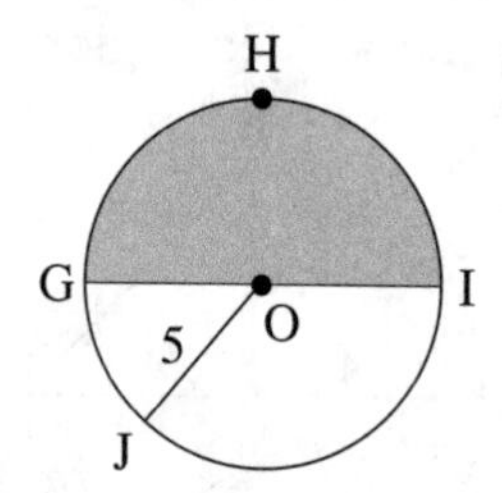

3. What is the diameter of a circle with area 36π?
4. What is the area of square *DEFG*?

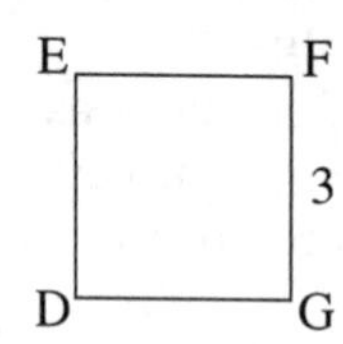

5. What is the area of rectangle *HIJK*?

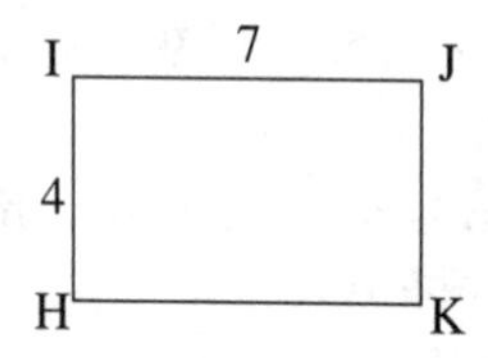

6. What is the area of ΔUFO?

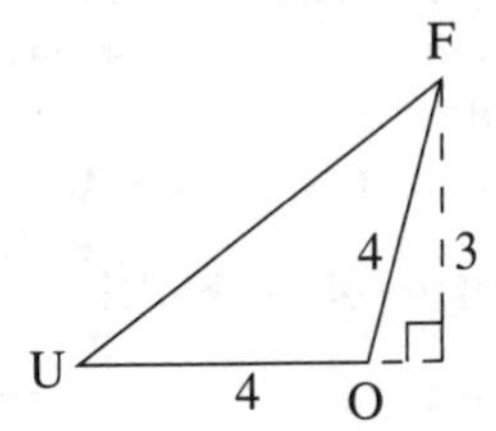

7. How many triangles congruent to ΔUFO in question 6 will fit into a rectangle with same dimensions as rectangle *HIJK* in question 5?

8. What is the measurement of QR in ΔPQS?

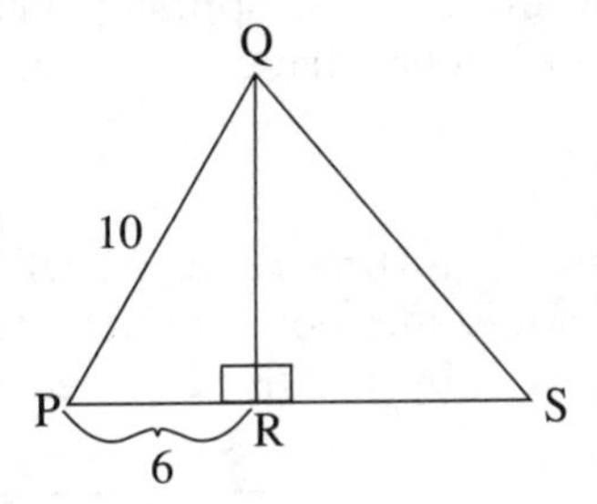

THREE-DIMENSIONAL SHAPES

Einstein said that time was the fourth dimension. And the Fifth Dimension was a band. If you don't remember that band part, ask someone about it. But before you get to the fourth dimension you have to deal with the other dimensions. Three-dimensional objects are objects that take up solid space, like pyramids, as opposed to two-dimensional objects that lie flat, like triangles. A book is three-dimensional, a page of a book is mainly two-dimensional.

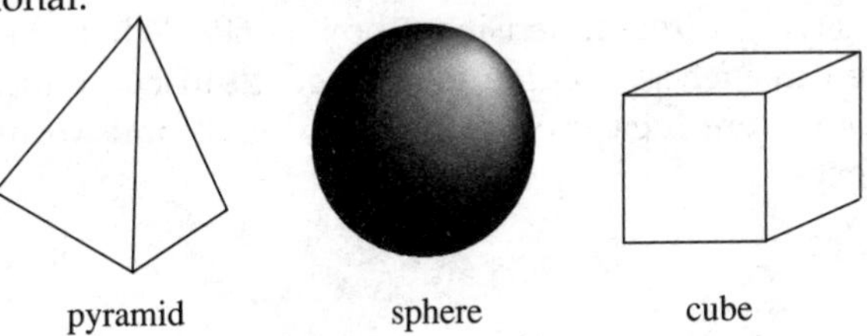

You could think about it as though the shapes we have been looking at are balloons, and now they are being blown up.

Whereas two-dimensional shapes have area, three-dimensional shapes have **volume**. Volume is the measurement of the space that a three-dimensional shape fills up. You know how you open a bag of potato chips and it is only $\frac{1}{4}$ full and you get steamed? Well, on the back of the bag says, "This package is sold by weight, not by *volume*," which means they don't care if it's filled up as long as it weighs 3 ounces or whatever.

Whereas two-dimensional shapes have perimeters or circumferences, three-dimensional shapes have **surface area**. Surface area is the area covered by all the surfaces of a solid shape. In other words, the amount of wrapping paper you would need to wrap a three-dimensional shape and cover all of its surfaces.

VOLUME

To find the volume of a three-dimensional object, treat it as though it were a two-dimensional object: find the area, and then multiply the area by the third dimension of the figure.

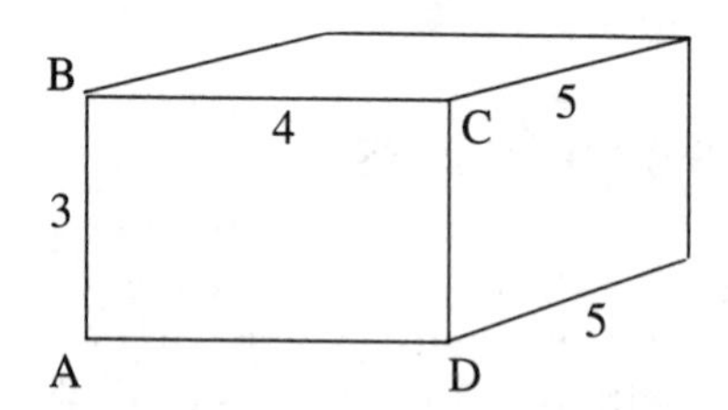

The area of rectangle *BACD* above is 12, because the length is 4 and the width is 3. And the other or "third" dimension is 5, so the volume of the **rectangular prism**, which is another way of saying three-dimensional box, is 60. When these units are given in inches, volume is referred to in cubic inches. So if this box were measured in inches, the volume would be 60 cubic inches.

Something to Think About

Why is it that area is measured in square inches and volume is measured in cubic inches, and these correspond to "square" and "cube"? A square inch is exactly that; it is a square with sides equal to 1 inch. The area of a 1-inch square is base times height, or 1×1, or 1^2.

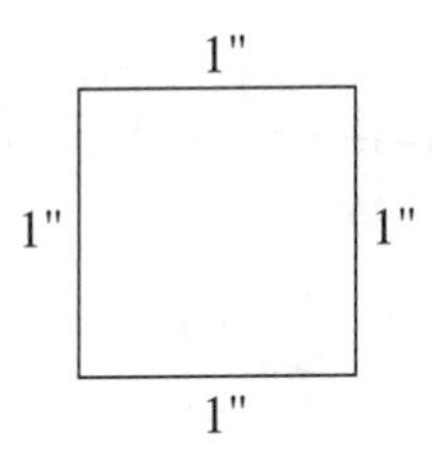

The area of any square is one of the sides, squared, because all the sides are equal. The volume of a three-dimensional square, or **cube**, 1 inch by 1 inch by 1 inch, is the area times the third dimension. Or, $1 \times 1 \times 1$ or 1^3. Remember 1^3 is read "one cubed." One cubic inch.

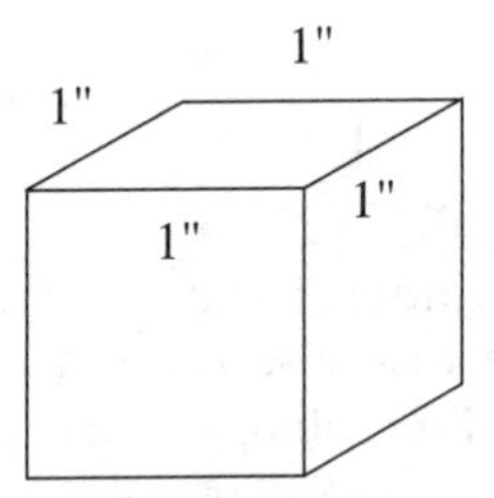

Volume of shapes represents how many 1 cubic inches will fit inside a three-dimensional object, like you are filling a box with sugar "cubes." However many sugar cubes fit inside a box is the box's volume.

TRIANGLES

What about with triangles? A **triangular prism**, that cool thing you can see light turn into rainbows through, is a long solid triangle. Like a Toblerone candy bar.

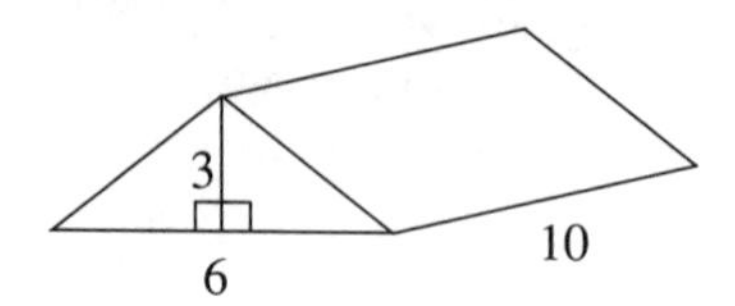

To calculate the volume, you need to find the area of the triangle part, and then multiply that by the third dimension, its length.

Area of a triangle is $\frac{1}{2}$ base times height.

$$\frac{1}{2} \times 6 = 3, \text{ and } 3 \times 3 = 9$$

The area of the triangle is 9, and the length of the prism is 10, so the volume of the prism is 9×10 or 90.

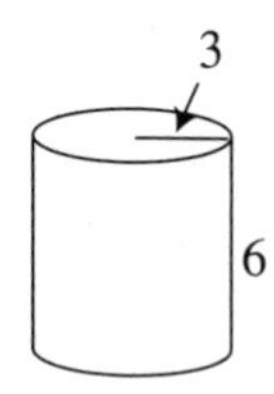

What about a **cylinder**? The volume of a cylinder is the area of the circle (in this case the radius is 3, so the area is 9π) times the third dimension, in this case the height, which is 6, so the volume is $6 \times 9\pi$ or 54π.

SURFACE AREA

The surface area of a three-dimensional shape is found by adding the areas of all the **faces** of the shape.

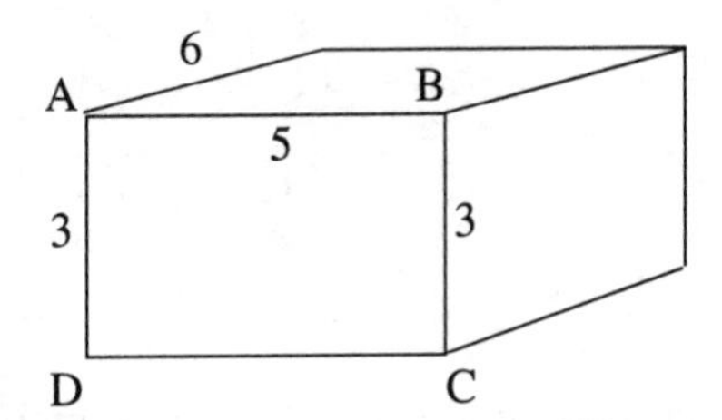

To find the surface area of a rectangular box or prism, like the one above, try thinking about unfolding the box so it is a flat rectangle with two flaps.

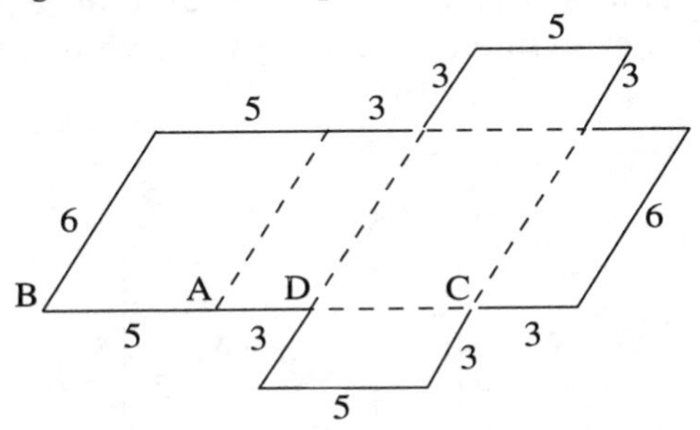

Now find the area of the big flat rectangle by adding up the lengths to form one big length.

$$5 + 3 + 5 + 3 = 16$$

The length is 16, and the width is 6, so the big rectangle has an area of 96. But you can't forget the flaps. Their width is 3, and their length is 5, so the area of each of them is 15. Two of them add up to 30, so the surface area of the whole is $96 + 30 = 126$.

The surface area is 126 square inches, assuming these measurements were inches. Why square? Because you are taking area, covering it with flat wrapping paper, right? Not taking up space, like volume.

A cylinder has surface area, too, and it's easy to see how it works.

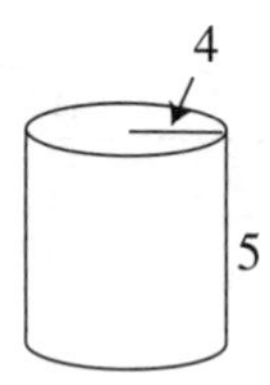

First find the area of the top and bottom of the cylinder. In this case, the radius is 4, so the area is 16π. Since there is a top and a bottom circle, you have $2(16\pi)$ or 32π. Now the side. Think of it as a soup can with a label. When you peel off the label all in one piece, not an easy thing to do, what shape is it?

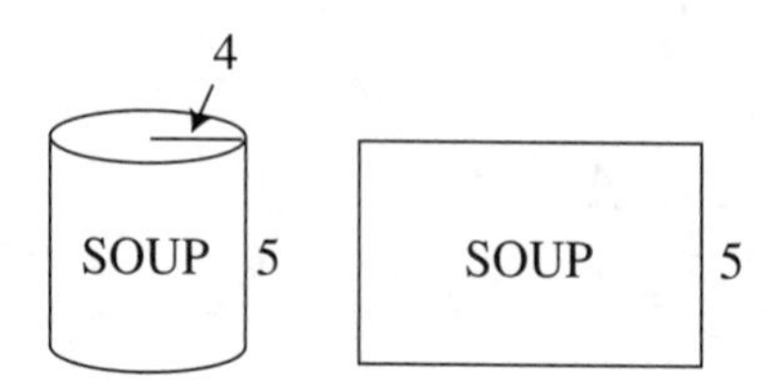

A rectangle. And the area of a rectangle is length times width. Well the width of this is given as the height of the cylinder, or 5. And the length is the edge around the top of the can, or the circumference. In this case, the circumference is 8π (radius is 4 so the diameter is 8, and circumference is diameter times π). The area of this rectangle is $8\pi \times 5 = 40\pi$. Together with the top and bottom of the cylinder, 32π, the surface is 72π square units.

GLOSSARY

Cartesian grid: A graph made up of a vertical (*Y*) and a horizontal (*X*) axis.
quadrant: A quarter section of a Cartesian grid corresponding to a change in negative or positive coordinates.
origin: The point where the *X* and *Y* axes of a Cartesian grid meet, at 0,0.
coordinates: Numbers assigned to a point on a Cartesian grid, also known as ordered pairs.
linear equation: An equation that expresses a line by using variables raised to the first power.
slope: The steepness of a line, expressed as the ratio of rise over run, also known as *m*.
rise: The change in a line along the *Y* axis of a Cartesian grid.
run: The change in a line along the *X* axis of a Cartesian grid.
vertex: The point where two lines meet to form an angle.
acute: An angle less than 90°.
obtuse: An angle less than 180° and more than 90°.
right: An angle measuring exactly 90°.
vertical angles: Opposite angles formed by the intersection of two lines.
supplementary angles: Angles that together add up to 180°.
complementary angles: Angles that together add up to 90°.
perpendicular: Lines that meet to form a right angle.
parallel lines: Lines that run equidistant from one another and never meet.
exterior angle: The angle formed by the extension of a line outside a shape.
triangle: A polygon having exactly three sides.
isosceles triangle: A triangle in which two sides are equal, and their corresponding opposite angles are also equal.

equilateral triangle: A triangle in which all sides and all angles are equal.
scalene triangle: A triangle in which no sides and no angles are equal.
right triangle: A triangle containing a right angle.
legs: Two sides of a right triangle that are not opposite the right angle.
hypotenuse: The side of a right triangle that is opposite the right angle.
Pythagorean theorem: The sum squares of the two legs of a right triangle will add up to equal the square of the hypotenuse.
congruent: Equal.
similar: Having the same ratio of sides and angles.
perimeter: The outline measurement of a figure.
area: The space covered within the outline of a figure.
base: The line to which a perpendicular line representing the height is drawn.
height: The perpendicular line from the highest point on a figure to its base.
quadrilateral: A polygon having four sides.
trapezoid: A quadrilateral with two parallel sides and two non-parallel sides.
parallelogram: A quadrilateral with two pairs of parallel sides.
rhombus: A parallelogram with four equal sides.
rectangle: A parallelogram with four right angles.
square: A rectangle with four equal sides.
length: The longer side of a quadrilateral.
width: The shorter side of a quadrilateral.
polygon: A closed plane figure of three or more sides.
radius: A line from the center of a circle to the edge.
diameter: A line from one edge of a circle to another, drawn through the center.
circumference: The perimeter of a circle.
π: Also known as "Pi", and pronounced "Pie." The ratio of the circumference to the diameter of a circle.
chord: Any line from one edge of a circle to another, that is not drawn through the center.
semicircle: One half of a circle.

arc: A section of the circumference of a circle. A minor arc corresponds to an angle less than $180°$.

sector: An arc plus the area its edges, drawn to the center, enclose. A minor sector corresponds to an angle less than $180°$.

tangent: A line, external to a circle, that touches one point on the edge of a circle.

volume: The amount of three-dimensional space a three-dimensional shape takes up.

surface area: The area covered by the external faces of a three-dimensional shape.

cube: A three-dimensional box in which all edges and faces are equal.

face: A side of a three-dimensional shape.

Now put it all together with a few geometry questions:

Quiz #6

1. Plot the line for the equation $y = 3x + 1$. What is its slope?
2. There is a sailboat with sail *DEF*. What is the measure of angle *EDF*? (We know you recognize that under all this fun wording of sailboats you recognize a geometry problem.)

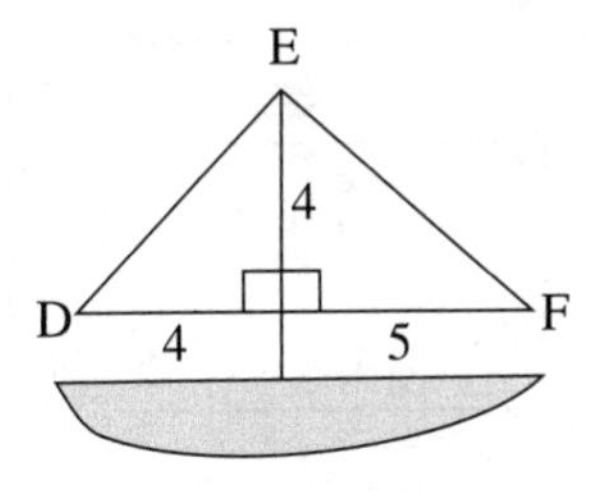

3. What is the area of the sail in question 2?

4. In parallelogram *ABCD,* what is the measurement of exterior angle *BAE*?

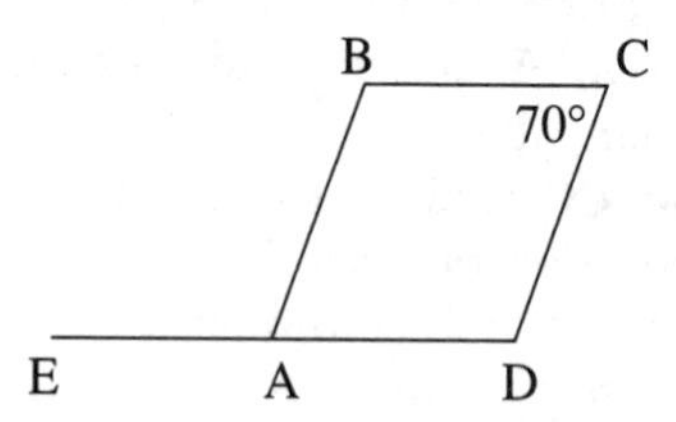

5. There is a pizza in which the side of each slice is 5 inches long. All slices meet at the center. What is the area of the pizza? What is its circumference?
6. What is the area of the shaded region in the square *GHIJ*?

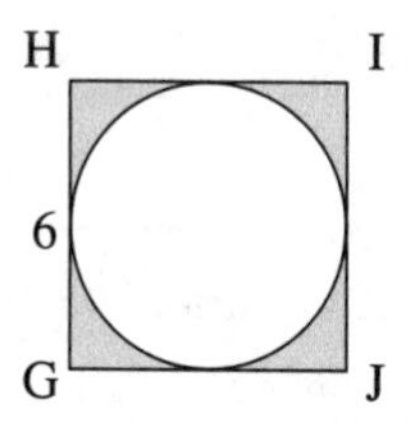

7. This is actually a box you can take to your favorite store and cram full of anything you want. For free. How much stuff will fit inside this box? In other words, what is its volume?

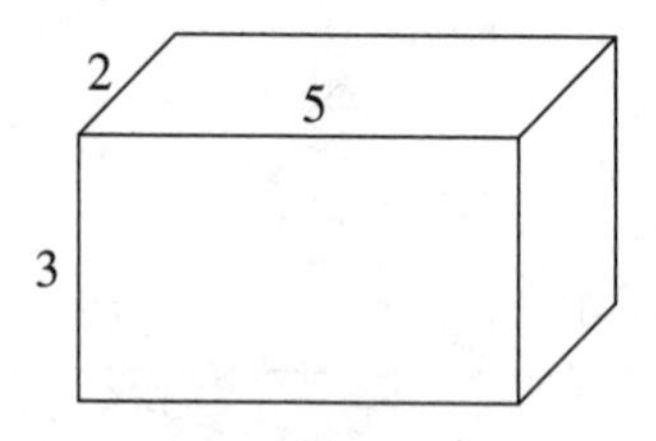

8. If you wanted to gift wrap the box in question 7 and give it to someone you loved, how much wrapping paper would you need (what is its surface area)?

9. This is actually a flat apple pie. You are being overwhelmed with a happy and healthy appetite, and taking that minor sector piece marked *DEF*. What is the area of the piece you are taking?

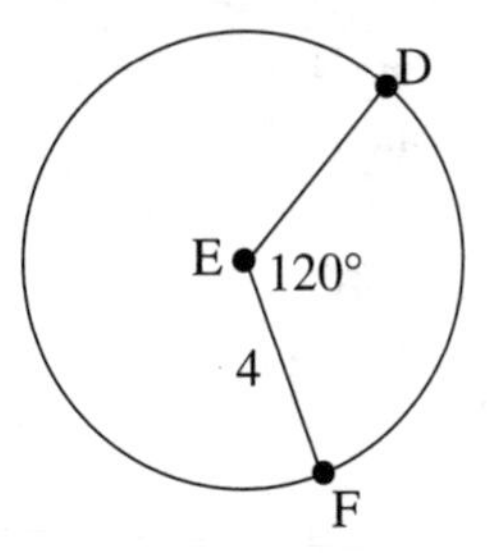

10. This is actually—see how your eyes can deceive you?—a girl leaning against a wall stretching her hamstrings before going out running. How tall is she, stretched out arm included, if her foot is 3 feet out from the wall and her hand is 5 feet up the wall?

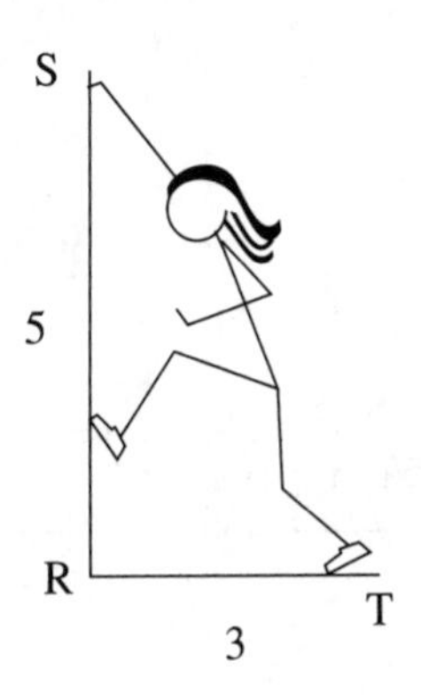

ANSWER KEY

Quiz #1

$A = (4,2)$ Quadrant 1
$B = (2,-3)$ Quadrant 4
$C = (-2,2)$ Quadrant 2
$D = (-5,6)$ Quadrant 2
$E = (-5,-1)$ Quadrant 3
$F = (0,0)$ Origin
$G = (0,7)$ Y axis
$H = (-6,0)$ X axis

Quiz #2

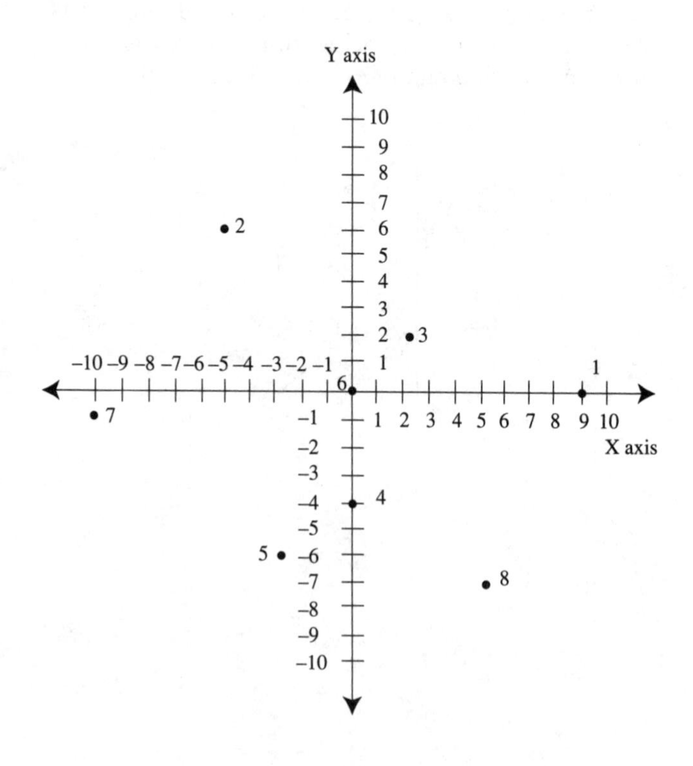

Quiz #3

1. a.

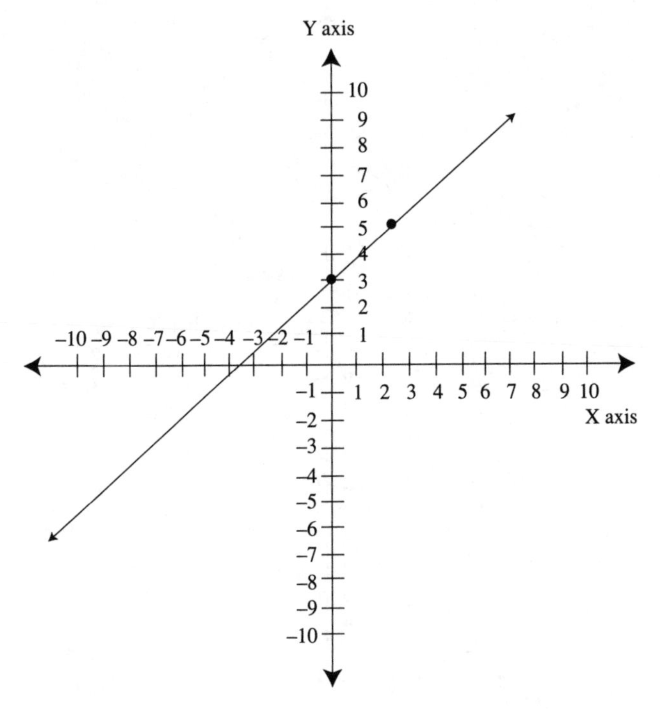

b. m (or slope) $= \frac{2}{2} = 1$

2. a.

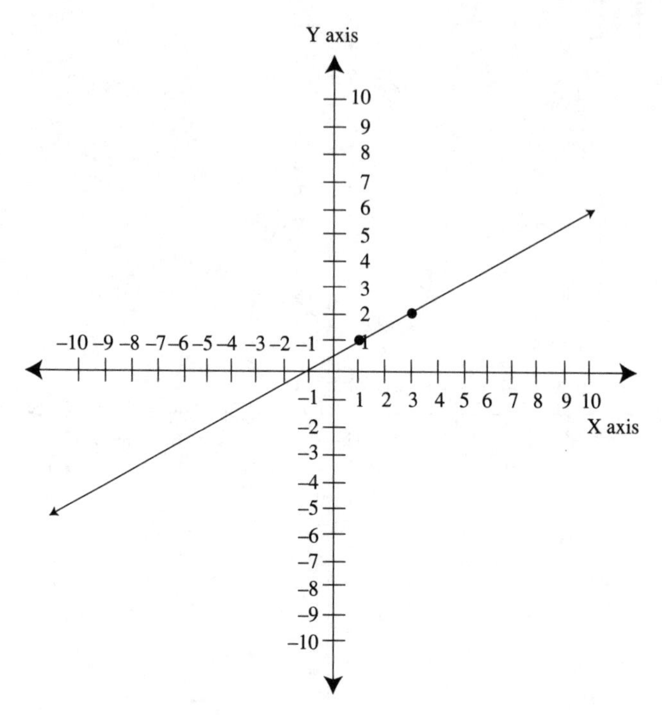
Y axis
10
9
8
7
6
5
4
3
2
1
–10 –9 –8 –7 –6 –5 –4 –3 –2 –1
–1
–2
–3
–4
–5
–6
–7
–8
–9
–10
1 2 3 4 5 6 7 8 9 10
X axis

b. $m = \frac{1}{2}$

3. a.

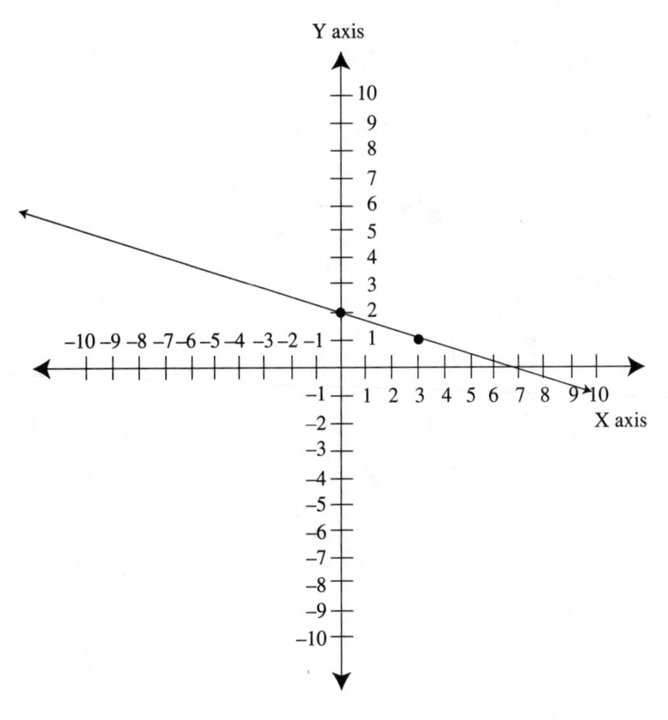

b. $m = -\frac{1}{3}$ (a negative slope slants the other way)

4. a.

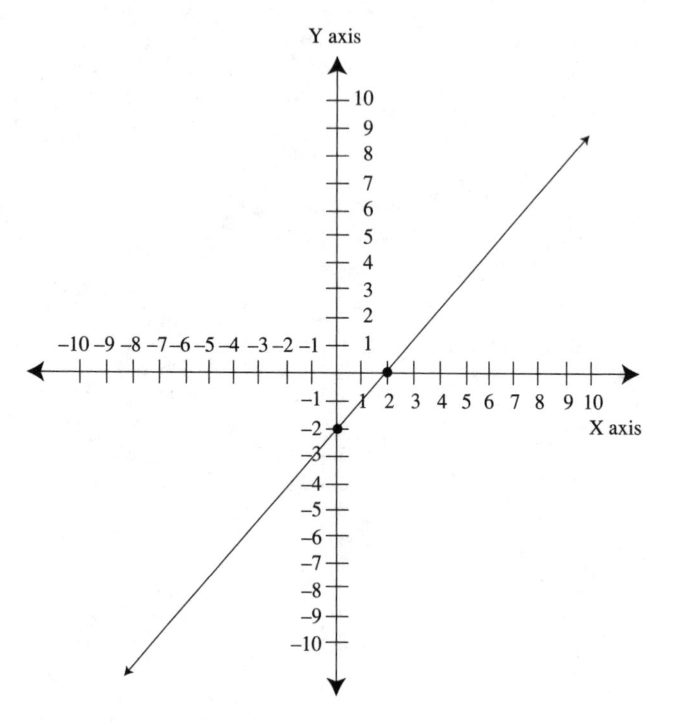

b. $m = \frac{2}{2} = 1$. Check it out, even though the line is on a different place on the graph than in number 1, the steepness of the slant is the same so the slope is the same.

Quiz #4

1. 30°
2. 15° each, remember? Exactly equal.
3. 12, or $\frac{1}{2}(4) \times 6$
4. 60° (supplementary angles)

How It Works:

If $\angle QSR$ is 20°, and $\angle SQR$ is 40°, the remaining angle in ΔQSR must be 120° to add up to 180°. Therefore, $\angle SRZ$ is supplementary to it, and must be 60° to add up to the measure of the line, 180°.

5. $13+\sqrt{13}$
6. $10\frac{1}{2}$ or $\frac{21}{2}$
7. Yes they are similar. No they are not congruent.
8. $12\frac{1}{2}$

How It Works:

If you turn the triangle upside down, you see that WX and XZ are meeting at a right angle, which means they are the base and the height. Area is $\frac{1}{2}$ base times height, or $\frac{1}{2}(5)\times 5$.

Quiz #5

1. 8
2. $\frac{25\pi}{2}$ which is $\frac{1}{2}$ the area of the circle, $\pi 25$
3. 12
4. 9
5. 28
6. 6
7. $4\frac{2}{3}$. Each one takes up 6, and it is really a division problem, "How many will fit into . . . ?" Divide 28 by 6 and get 4 triangles and then 4 spaces remaining or $\frac{4}{6}=\frac{2}{3}$.
8. 8

Quiz #6

1.

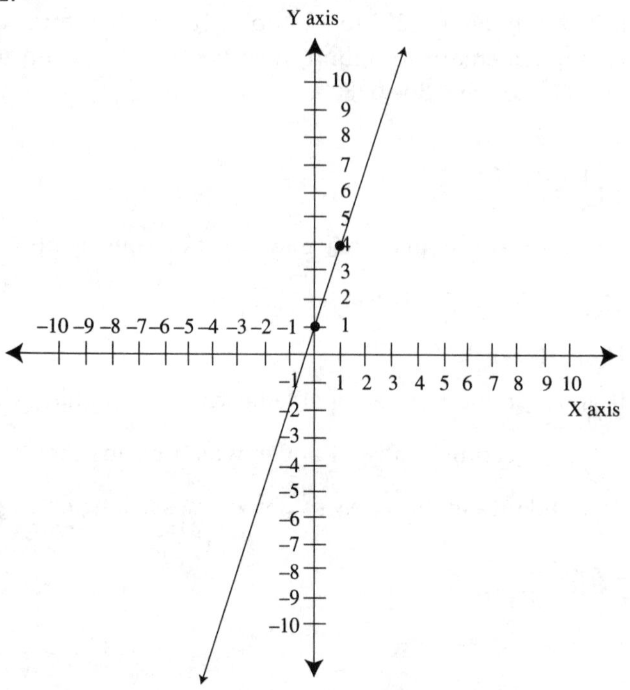

$m = 3$

How It Works:

Try 2 values for x. The first is 0, and that means $y = 1$. The second is $x = 1$, and $y = 4$. Then try a third to test. $x = 2,\ y = 7$. To find the slope, $\frac{y_2 - y_1}{x_2 - x_1} = \frac{4-1}{1-0} = \frac{3}{1} = 3$.

2. 45°

How It Works:

Since the sail is made up of two smaller triangles, you can look at the left triangle to find the measure of this angle. It has two equal sides, 4 and 4. And the angle that is not equal is 90°. That means you have an isosceles triangle with a right angle, so the two equal angles will add up to 90°.

$\frac{90}{2} = 45$. Both $\angle EDC$ and $\angle DEC$ are 45°.

3. 18

How It Works:

Area is $\frac{1}{2}$ base times height. In this case, the base of the whole sail is 9, and the height is $4.9 \times 4 = 36$

$$\frac{1}{2} \times 36 = 18$$

4. 110°

How It Works:

The exterior angle here is supplementary to $\angle BAD$. And $\angle BAD$ is diagonally across from angle *BCD*, which is 70°. This means $\angle BAD$ and $\angle BCD$ are equal. Remember, opposite angles in a parallelogram are equal? And $\angle BAE$ is supplementary to $\angle BAD$, or 70°, they need to add up to 180°, so $\angle BAE$ must be 110°.

5. area $= \pi 25$; circumference $= \pi 10$

How It Works:

A side of a slice is 5, and that also means the radius is 5, because each side of a slice goes from the center to the edge. Since the radius is 5, the area is πr^2 or $\pi 25$. The circumference is π times the diameter or $\pi 10$.

6. area $= 36 - 9\pi$

How It Works:

The area of the shaded region is the area of the bigger shape, the square, minus the area of the smaller shape, the circle. The square has an area of 36, because one side is 6. For the area of the circle you need the radius. Well, take a look. The side of the square is 6, and if you look at the circle you see that the diameter is equal to the side of the square. The diameter of the circle then, is 6, so the radius is 3, or $\frac{1}{2}$ of 6. The area of a circle with radius 3 is $\pi 9$. The shaded region is $36 - \pi 9$.

7. 30

How It Works:

The volume of a three-dimensional figure is found by getting the area of a face and then multiplying the area by the other dimension. The front face here has an area of 15, because length, 5, times width, 3, equals 15. The other dimension is 2, and 15 times 2 equals 30.

8. 62

How It Works:

Unfold the box; make it more visual by drawing a picture. Or, if you have a hard time visualizing this, take the area of each separate face and add them all together. Unfolded, the length of the box is 16, and the width is 2.

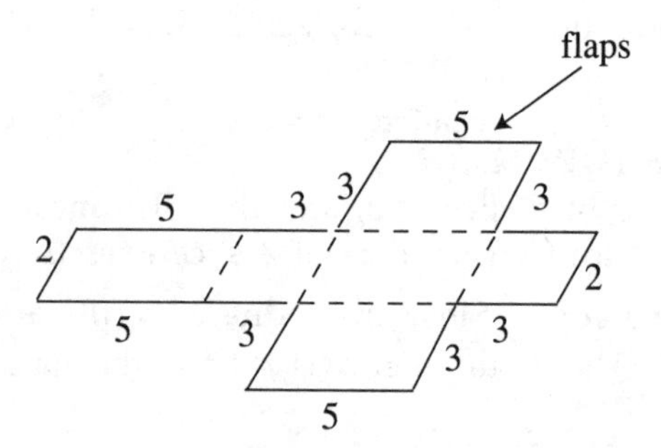

So the area of the box without flaps unfolded is 16 times 2, or 32. Now add in the area of the two flaps with length 5 and width 3. They each have an area of 15; together they have area 30. $30 + 32 = 62$. The surface area is 62.

9. $\frac{\pi 16}{3}$

How It Works:

The piece of pie is really a sector, and to find the area of the sector you want to find the area of the whole pie. The radius of the pie is 4, and area is π (sorry, we really don't intend the pun) r^2. The area of the pie is $\pi 16$. The sector has an angle measure of 120°. What part of the whole is that? $\frac{120}{360} = \frac{1}{3}$. This means the piece is $\frac{1}{3}$ of the area of the whole. $\frac{1}{3} \times \pi 16 = \frac{\pi 16}{3}$. That is the area of the piece of pie.

10. $\sqrt{34}$ or just a bit under six feet tall.

How It Works:

This girl forms the hypotenuse of a right triangle, so you can use the Pythagorean theorem on her. The floor is 3 and the wall is 5.

$$3^2 + 5^2 = \text{girl}^2$$

$$9 + 25 = \text{girl}^2$$

$$34 = \text{girl}^2$$

$\sqrt{34} = \text{girl}$. You can approximate this, right? $\sqrt{36}$ is 6, and $\sqrt{34}$ is a little less than six, so she's a bit under six feet tall when she is stretched out with her arms up. Measure your own arms and figure out how tall she might really be, without her arms above her head.

CHAPTER 7

PROBABILITY AND STATISTICS

Probability is the mathematical measure of the likelihood that something will happen. It generally applies to the future. **Statistics** is used to measure the relative amounts and occurrences of different things. These things have usually already happened or are happening, so statistics applies mostly to the past or present.

PROBABILITY

How can you measure the likelihood that something will happen? The way to use math to express likelihood involves knowing what the possibilities are. For instance, if there were a box of marbles with 5 red marbles and 6 blue marbles, and you get to choose 1 marble without looking, what is the probability that it would be red? First, how many possible marble choices are there? There are 11 possibilities, because each marble has the same chance of being chosen. How many of those chosen could be red? Well there are 5 red marbles, so there are 5 red possibilities out of 11. You set this up in fraction form to express probability. The probability of choosing a red marble at random out of this particular box is $\frac{5}{11}$.

What happens when there are no red marbles in the box? For instance, if there were a box of 11 blue marbles, what is the probability of choosing a red marble? Set it up the same way. How many choices are there? There are 11. How many of these could be red? 0.

Set this up as a fraction: $\frac{0}{11}$. This gives you 0 probability. When you find a 0 probability for something, it means the event, in this case choosing a red marble, will definitely not occur.

What happens if there are all red marbles in the box? Same set-up. There are 11 marbles in another box, and every last one of the pesky marbles is red. What is the probability of your choosing a marble out of this box and its being red? How many possibilities? 11. How many red possibilities? 11. Probability? $\frac{11}{11}$, better known among the math smart like yourself as 1. When you find a probability of 1, it means the event, again in this case choosing a red marble, will definitely occur.

Try some probabilities now, before you take off for Vegas.

Quiz #1

1. What is the probability that a coin with one side heads and the other side tails, will turn up heads on one toss?
2. What is the probability of getting an orange Life Saver in a roll of Life Savers with 2 greens, 2 reds, 2 oranges, and 1 yellow?
3. What is the probability of getting a yellow Life Saver out of the same roll?
4. What is the probability of getting a red Life Saver out of the same roll?
5. What is the probability of getting a green Life Saver out of the same roll?
6. In a box with 4 black marbles, 3 white marbles, and 2 purple marbles, what is the probability of getting a purple marble when choosing only 1 marble, at random?
7. How about a black marble?
8. How about a white marble?

You are probably thinking that NOW you can go to Vegas because you know how to find probabilities of all sorts of different things. It's true, you do know different probabilities, but before you go you need to see how some more complex probabilities can be worked out. Besides, the probability of winning any money in Vegas is almost 0. And you know what that means.

COMBINATIONS

Combinations are just what they sound like, combinations of various things. For instance, if you were choosing 3 marbles out of a box of 4 marbles, 1 red, 1 yellow, 1 blue, 1 green, how many combinations could you pick? R is red, Y is yellow, B is blue, and G is green.

List them out.

RYB RYG RBG BYG

What about GYR? In combinations, the order doesn't matter. That means RYG and GRY are the same combination of colors. So there are only 4 different combinations of colors to choose from the box. This is useful when figuring out things like how many possible hands of poker can be dealt from a deck of cards, because it doesn't matter what order the cards come in, the only thing that matters is the combination of cards you are dealt.

How can you figure out combinations mathematically? Well it is always a good idea to list the possibilities if you are working with small amounts, like 3 or 4. And it is also good to take a look at the formula for combinations. But first you need to learn a new term. **Factorial**. The factorial of a number is that number times every positive whole number smaller than itself, down to 1.

The factorial of 5 is $5 \times 4 \times 3 \times 2 \times 1$. This equals 120.

The math symbol for factorials is really great. The way to write the factorial of 5 is 5! We know it makes you want to say 5! What a fabulous number! 5! I can't live without you oh wonderful 5! But the way to read it aloud is to look at 5! and say, "Five factorial." And just to keep you on your toes, by definition, 0! is equal to 1.

The formula for combinations is the factorial of the number of things being chosen from (in our—the situation where you are choosing 3 marbles from a box of 4—the number of marbles being chosen from is 4). This number is then divided by the factorial of the number of things chosen (3 for us) times the factorial of the difference between the two (4—3, or 1, in this case).

$$\frac{4!}{3!(4-3)!} = \frac{4 \times 3 \times 2 \times 1}{3 \times 2 \times 1(1)}$$

You can cancel here, can't you?

$$\frac{4 \times \cancel{3} \times \cancel{2} \times \cancel{1}}{\cancel{3} \times \cancel{2} \times \cancel{1} \times (\cancel{1})}$$

This equals 4. There are 4 combinations possible, but you already knew that, didn't you?

If it makes you more comfortable to look at the formula algebraically, here is what it looks like. First, combinations of things are expressed as C_r^n where n is the pool of things chosen from and r is the number of things chosen.

$$C_r^n = \frac{n!}{r!(n-r)!}$$

As long as you understand what the formula means, you are in good shape. Take a look at another combination before you go off and do one on your own.

There are 5 electives offered at your school: woodworking, sculpture, jazz band, drama, and photography, and you are supposed to choose 2 of them. How many possible combinations of the 2 are there?

You need the group, 5, factorial over the number chosen, 2, factorial times the difference between these, factorial.

$$\frac{5!}{2!(5-2)!} = \frac{5\times4\times3\times2\times1}{2\times1\times(3\times2\times1)}$$

You can cancel, right?

$$\frac{5\times\overset{2}{\cancel{4}}\times\cancel{3}\times\cancel{2}\times\cancel{1}}{2\times1\times(\cancel{3}\times\cancel{2}\times\cancel{1})}$$

There are 10 possible combinations. Pretty slick, no? You can also list the combinations.

WS WJ WD WP
SJ SD SP
JD JP
DP

See how starting with each elective creates one less combination as you go down? Consider that pattern; it shows up a lot in combinations.

APPROXIMATE THIS:

How many different possible 5-card poker hands can be dealt from a standard deck of 52 cards? Remember you don't have to work out the math here, just take a guess.

Can you see how this works into probability? Since there are 10 possible combinations of electives, and they are all equally likely, what is the probability that your friend Jenny is taking woodworking and drama (if it's random and she doesn't prefer one elective over another, and has no control of her own life)? Well that particular combination only comes up once, so the probability is $\frac{1}{10}$.

Try some combinations on your own. If the numbers are small enough, try listing the possibilities.

Quiz #2

1. How many possible combinations of toppings are there if you choose any 1 from the group of nuts, sprinkles, hot fudge, butterscotch, and marshmallow?
2. How many possible combinations if you choose any 2 from the group of toppings?
3. How many possible combinations if you choose any 3 from the group of toppings?
4. How many possible combinations if you choose any 4 from the group of toppings?
5. How many possible combinations if you choose any 5 from the group of toppings?
6. The band Babe the Blue Ox is choosing 3 of their 6 songs, "Elephant," "Spatula," "Chicken-head-bone-sucker," "Born Again," "Axle," and "Home," to appear on their new CD. How many possible combinations of songs are there?

7. Jane is choosing 2 essays from among 7 essays on politics, academics, sports, family, carpentry, history, and math, to fill out her college application. How many possible combinations are there?
8. Fred can choose 3 CDs from among 10 different CDs at his new CD-of-the-month club. How many possible combinations are there for him to choose from?

PERMUTATIONS

You know how to choose groups of things without regard to order, now you can try arranging possibilities when order is important.

Permutations are arrangements of things in which order matters. For instance, in a road race how many combinations of 3 runners, Sally, Mark, and Elliot, are there? If only the 3 run, there is only 1 combination, Sally, Mark, and Elliot in whatever order. But the order does matter, so you need to find how many permutations there are. To figure permutations you get to use factorials! again. The way to calculate a permutation is to put the factorial of the pool on top—in this case 3—and then the factorial of the difference between the pool and the number being selected—in this case 3—3 or 0.

The algebraic expression, where you use P as permutation, again you use n as the pool of objects, and r as the number of objects chosen, is $P_r^n = \frac{n!}{(n-r)!}$.

$$\frac{3!}{(3-3)!} = \frac{3!}{1}. \quad \text{Remember, } 0! = 1.$$

$\frac{3 \times 2 \times 1}{1} = 6$. There are 6 different orders in which these 3 people can finish the race. What are they? List them.

SME SEM MES MSE ESM EMS

Take a look at another permutation. The 3 most politically aware senators are being chosen from among a group of 5 senators, Adams, Boyd, Carlton, Demmings, and Eaton, that are in a special club of politically aware people. Since you are choosing the most politically aware, and later you are going to post their rankings in their states, order will be crucial. How many possible orderings will there be?

Well the total group, which is 5, goes on top, and the difference between the group and the number of people chosen goes below.

$$\frac{5!}{(5-3)!} = \frac{5 \times 4 \times 3 \times 2 \times 1}{2 \times 1} = 60$$

That's a lot, isn't it? Maybe a lot more than you thought? There are always going to be at least as many permutation possibilities as combination possibilities, and probably more, because the order allows for more possibilities. For instance, a two layer casserole, this might not sound tasty to you, of turkey and cheese has only one possible combination of ingredients, but two possible permutations, turkey on top and cheese on the bottom or turkey on the bottom and cheese on top.

Figure the following permutations:

Quiz #3

1. You are ordering an ice cream cake made in layers. How many possible ways are there to order a 4-layer cake with 4 flavors of ice cream, chocolate, vanilla, strawberry, and mocha? Assume no flavor is repeated.
2. How many possible ways are there to order a 3-layer cake out of the above 4 flavors of ice cream?
3. How many possible ways are there to order a 2-layer cake from among the 4 flavors of ice cream?
4. How many ways are there to order a 1-layer cake from among the 4 flavors of ice cream?
5. In the race for president, there are 7 candidates. If only the first 3 get ranked, how many possible ways are there to order the 3 ranked candidates?
6. In the special prize for best book, only the top 2 books of the 6 bestsellers, titled *Grandeur, Pride, Envy, Lust, Blood,* and *Power* get any prize money, with

the top book getting $2,000 and the second book getting $1,000. How many possible orders of books to get prize money are there?

7. A museum has 5 pictures to arrange on the wall. How many possible arrangements are there? The order matters; remember their artistic vision.
8. What if they thought 5 paintings was excessive and embarrassing, how many possible arrangements would there be if they decided to select 4 out of the 5? The order of the paintings still matters.

It is always a good idea to list possibilities when there aren't too many of them. It will give you a nice solid tangible piece of information to work from, and that is about as math smart as you can get. The more you understand what numbers and expressions are trying to represent, the better off you are.

Permutations can help you understand more sophisticated ways of gauging probability. When you arrange permutations, you use the factorial. A record with 5 songs could be ordered $\frac{5 \times 4 \times 3 \times 2 \times 1}{1}$ different ways. This is because there are 5 possibilities for the first song. Once you have placed the first song, there are 4 remaining possibilities for the second song, and so on. You can use this idea when calculating how many possibilities there are when repeating something like flipping a coin or throwing a pair of dice.

How many possibilities are there when flipping a coin once? Two possibilities: heads or tails.

What if you flip a coin twice? There are 4 possibilities, heads tails, tails heads, heads heads, or tails tails. From now on we will use H for heads and T for tails.

To figure the possibilities, multiply the possibility of each time, because unlike ordering the songs on the record, your possibilities do not decrease with each successive toss. $2 \times 2 = 4$.

Flipping a coin 3 times will give you how many possibilities? $2 \times 2 \times 2 = 8$.

List them out: HHH, HHT, HTH, HTT, TTT, TTH, THT, THH.

It works with dice, too. If you roll a 6-sided die, each face of the die has a different number 1 through 6, how many possibilities are there with one roll? 6. With two rolls? Multiplying 6 times 6 equals 36, since the second roll does not have fewer possibilities. What is the probability of rolling snake eyes—a pair of 1s—if you roll a pair of 6-sided dice?

Well, there are 36 possibilities, as you just saw. Only one of these possibilities has a pair of 1s, so the probability is the number of possible successes, or 1, over the number of possibilities total, or 36. The probability of rolling snake eyes is $\frac{1}{36}$. Now you can see why some games have crazy things like a 12-sided die; these games alter the probability of rolling various numbers.

STATISTICS

Statistics, as you read before, measures relative occurrences. How does it differ from probability? Probability measures the likelihood that something will happen. Statistics is a way of organizing mathematical information.

The way you have seen statistics used most often is probably **averages**. For instance, you might say, "Oh I did about average on the French test." What does this mean?

The average of a group of numbers, called a **sample** in statistics, is found by adding all the numbers in the sample together, and then dividing by how many numbers there are. Average is also called the **mean** or the **arithmetic mean** by some math people, just for a little more fun and variety.

If, on the French test, one person got a 78, you got an 84, and the only other person in the class got a 90, did you really do about average?

To find out, first take the sum of the 3 scores. $78 + 84 + 90 = 252$

Now, divide the sum by the number of scores you are averaging. How many scores? 3.

$$\frac{252}{3} = 84$$

The average score of this little class is an 84. That means your score was exactly average. And you notice that in this case, average doesn't mean a C, does it. It just means the sum divided by the number in the sample.

$$\text{Average} = \frac{\text{sum}}{\text{number in sample}}$$

This means that if you have any two of these elements—average, sum, or number in sample—you can figure out the other.

If the average weight of 5 people is 140, how much do they weigh all together? You have the average and you have the number in the sample, all you are missing is the sum.

$$140 = \frac{\text{sum}}{5}$$

You can manipulate your equation now, just like in algebra. Multiply both sides by 5.

$700 = \text{sum}$. You now know how much everybody weighs all together.

Something to Consider for a Few Moments

Does this mean you know how much any one particular person weighs in the group? No. You only have the average weight of each, so you have no true idea of how much any one person weighs except that no one weighs more than 700 pounds because that would exceed the sum. But it is theoretically possible for one person to weigh exactly 700 pounds, and everybody else to weigh 0, and still get an average weight per person of 140. For this reason, beware of people who try to make what sound to you like unreasonable points with statistics. Statistics can illuminate information, but they can also obscure it, and they must be looked at carefully with a healthy mathematical skepticism. The way averages can cover up extreme numbers also causes some people who calculate averages to drop both the lowest and highest numbers in a sample in the hope of getting a more representative average.

APPROXIMATE THIS:

How many different possible ways are there to make the first ten moves in a chess game?

MEDIAN

There are other statistical measures that use samples, that some people confuse with averages. The first is called the **median.** The median of a sample is the number that occurs exactly in the middle. For instance, there are 7 kids waiting to get in to see a NC-17 movie. Their ages are 15, 15, 16, 17, 18, 18, 20. The median of their ages is 17, because 17 is the number exactly in the middle. Just count from the outsides until you get to the exact middle number of the group. If a sample is not in numerical order, put it into numerical order and then count to the number in the middle. If there are an even number of items in a sample, your method is a little bit different.

There are 4 CDs in a sale box. One of them has 3 songs, one has 15 songs, one has 4 songs, and one has 2 songs. What is the median number of songs per CD in the box?

Well first you need to put the numbers in order.

2, 3, 4, 15

Now, to find the middle number, look at the two middle numbers, 3 and 4. You need their average. The average of the two middle numbers is counted as the median of an even-numbered sample.

$$\text{Average} = \frac{\text{sum}}{\text{number in sample}}$$

$$\text{Average} = \frac{7}{2}$$

$\text{Average} = 3.5$. So the median number of songs per CD in the box is 3.5. Notice how the median is different from the average number of songs per CD. The average is the sum divided by the number. In this case, the average is $\frac{24}{4}$ or 6 songs per CD.

Think about how you might choose to use either number if you were an unscrupulous salesperson. Which would you use? If this interests you, look for a book called *How To Lie With Statistics*, by Darrell Huff. It is filled with information like this.

MODE

The **mode** of a sample is that number that occurs most frequently within a sample. For instance, if Sharon were to measure the height of every person in her family, her sample might look like this.

Mother, 5′ 2″. Father, 5′ 10″. Older sister, 4′ 11″; middle sister, 5′ 4″; Sharon, 5′ 2″. Does any number occur more than once? Only one number, 5′ 2″. This means that 5′ 2″ is the mode of Sharon's family height sample, because no other number appears that many times. Figure out what the mode of your family sample is.

If no number occurs more than once, there is no mode.

What if two different numbers occur the same number of times?

Lisa's family: Lisa, 5′ 6″. Her mother, 5′ 6″. Her father, 6′. Her brother David, 6′. Her brother John, 5′ 11″. Her brother Bob, 5′ 10″.

Both 5′ 6″ and 6′ appear twice, and they both appear more than the other heights. You can call Lisa's family a **bimodal sample**, because there are two modes, 5′ 6″ and 6′.

APPROXIMATE THIS:

What is the average height of a U.S. man? Of a U.S. woman?

GLOSSARY

probability: The mathematical likelihood that a specific event will happen.
statistics: A mathematical way of analyzing information.
combinations: Possible combinations of things or events without regard to order.
factorial: The product of a number multiplied by every positive integer less than itself.
permutations: Possible combinations of things or events with regard to order.
average: Also known as mean or arithmetic mean, the number that represents the norm of a group of numbers, by dividing the sum of the numbers in a sample by the quantity of numbers in the sample.
sample: A pool of numbers being analyzed.
mean: The average.
median: The number that occurs directly in the middle of a sample, numerically.
mode: The number that occurs most frequently in a sample.

Try the last (can you believe it?) set of chapter problems. Take your time, stretch it out, think of ways you can use statistics to fool people into believing crazy things.

Quiz #4

1. You have a 6-sided die, each face has one number from 1 through 6. You roll once. What is the probability that you roll a 6?
2. How many different ways are there to make a 2-layer cake out of 3 different layers, lemon, angel food, and devil's food? Assume that the order of the layers does not matter.

3. There are 10 people trying out for the basketball team, Sue, Mike, Grace, Jeff, Alice, John, Erica, George, Berna, and Perry. Only 7 will be chosen. How many possible combinations of the 7 are there?
4. What is the probability that a card chosen at random from a standard deck of 52 cards is an 8?
5. Joan's compact disc player has a setting called "random" that plays the songs on a particular compact disc in any random order. If she is playing a compact disc with 7 songs, what is the probability that the order of the songs will be the exact same order as the order given on the compact disc (1-7)?
6. If you ate 1 carrot on Monday, 2 carrots on Tuesday, 1 carrot on Wednesday, 3 carrots on Thursday, 7 carrots on Friday, 5 carrots on Saturday, and, mercifully putting an end to the madness, 4 carrots on Sunday, what was the average number of carrots you ate per day on this specific week?
7. What was the median number of carrots you ate during that hellish week?
8. What was the mode of the sample of carrots you ate that week in question 6?
9. A group of students together received money totaling $288, with each student getting an average payment of $36. How many students were there in this specific group?
10. A group of 5 girls took 120 minutes (total) to do a math problem. A group of 3 boys took 81 minutes (total) to do the same problem. What is the difference between the average time it took a girl and the average time it took a boy to do this problem, and who did it faster?

ANSWER KEY

Quiz #1

1. $\frac{1}{2}$
2. $\frac{2}{7}$
3. $\frac{1}{7}$
4. $\frac{2}{7}$
5. $\frac{2}{7}$
6. $\frac{2}{9}$
7. $\frac{4}{9}$
8. $\frac{3}{9}$ or $\frac{1}{3}$ (yes, you can reduce it!)

Quiz #2

1. nuts, sprinkles, fudge, butterscotch, or marshmallow.
   5 possibilities.
2. NS, NF, NB, NM
   SF, SB, SM
   FB, FM
   BM
   10 possibilities.
3. NSF, NSB, NSM
   NFB, NFM
   NBM
   SFB, SFM
   SBM
   FBM
   10 possibilities.
4. NSFB, NSFM, NSBM
   NFBM, FSBM
   5 possibilities.
5. NSFBM
   1 possibility.
6. 20 possibilities; you can try to list them all.
7. 21 possibilities; you can list these, too.
8. 120 possibilities; don't bother with the list if you don't want to get a serious hand cramp.

Quiz #3

1. 24 possible ways.
2. 24 possible ways.
3. 12 possible ways.
4. 4 possible ways, one cake of each flavor.
5. Can you believe this? There are 210 possible ways for those first three candidates to get ranked.
6. There are 30 ways to rank these top two books.
7. There are 120 different ways to order the paintings.
8. There are also 120 different ways to order 4 out of 5 of the paintings.

Quiz #4

1. $\frac{1}{6}$

   How It Works:

   Probability is the number of successful possibilities over the number of possibilities total. With this die there are 6 different possibilities on a roll, and only 1 successful (to the question) roll of 6 possible. This means the probability is $\frac{1}{6}$.

2. 3 possible ways

   How It Works:

   The possible cakes are LA, LD, and AD. The way to look at this with a formula is to set up the formula for a combination, since order does not matter according to the question.

   $$C_3^2 = \frac{3!}{2!(3-2)!} = 3 \times 2 \times \frac{1}{2} \times$$

   $$1(1) = 3.$$

3. 120 possibilities

   How It Works:

   This is also a combination problem, because the order of the people chosen for the team does not matter.

   $$C_{10}^{7} = \frac{10!}{7!(10-7)!} =$$

   $$\frac{10 \times 9 \times 8 \times 7 \times 6 \times 5 \times 4 \times 3 \times 2 \times 1}{7 \times 6 \times 5 \times 4 \times 3 \times 2 \times 1(3 \times 2 \times 1)}$$

   $$10 \times 3 \times 4 = 120$$

4. $\frac{4}{52}$

   How It Works:

   How many successful possible draws are there? There are four 8s in a deck with 52 cards. So there are 4 successful possibilities out of 52 possible draws. This means the probability of an eight is $\frac{4}{52}$.

5. $\frac{1}{5,040}$

How It Works:

The possible orders, since order matters here, can be figured out by permutation. You choose 7 out of 7, so your calculation looks like this.

$$P_7^7 = \frac{7!}{(7-7)!} =$$

$$\frac{7 \times 6 \times 5 \times 4 \times 3 \times 2 \times 1}{1} =$$

$$5{,}040$$

Probability refers to how many successful possibilities there are. Only one order will go exactly song 1 through song 7. This means, $\frac{1}{5,040}$ is the probability of getting 1 through 7. Listen and see if it ever happens on a CD you are listening to on "random."

6. $\frac{23}{7}$ or $3\frac{2}{7}$

How It Works:

The average is the sum divided by the number in the sample. The sum of the carrots eaten in this particular week is $1+2+1+3+7+5+4=23$. The number in the sample, in this case number of days in a week, is 7. The average is $\frac{23}{7}$ or $3\frac{2}{7}$ carrots per day.

7. 3

How It Works:

The median is the exact middle number of a sample. You need to line up all the numbers in size order, and then count. 1, 1, 2, 3, 4, 5, 7. The middle number is 3, so 3 is the median of this sample.

8. 1

How It Works:

The mode, as we are sure you remember, is that number in the sample that occurs most frequently. In this sample, 1, 2, 1, 3, 7, 5, 4, the mode is 1 because 1 is the only number that occurs more than once.

9. 8

How It Works:

You are being asked here to find the number of students in the sample. You are given the average and the sum of the money. You can set up the formula and take a look.

$$36 \text{ (average)} = \frac{288\text{(sum)}}{?\text{(number in sample)}}$$

$36 = \frac{288}{?}$. To find the missing information you can manipulate the equation by multiplying both sides by the unknown. $36 \times ? = 288$. Now divide both sides by 36 so you have the unknown amount isolated.

$? = \frac{288}{36} = \frac{72}{9} = 8$. There are 8 students in the sample.

10. The girls took an average of 24 minutes to do the problem, the boys took an average of 27 minutes to do the problem, therefore the girls, on average, were about 3 minutes faster.

How It Works:

To find the average time per problem of the girls, divide the total time by the number of girls in the sample.

$$\text{Average} = \frac{120}{5} = 24$$

To find the average time per problem for the boys, again divide the total by the number of boys.

$$\text{Average} = \frac{81}{3} = 27$$

The boys, on average, took 3 minutes more to do the problem than the girls, so the girls did it 3 minutes faster.

APPROXIMATION ANSWERS

ANSWER KEY

APPROXIMATE THIS

CHAPTER 1

1. About 103,524, and exactly 103,523.
2. The earth weighs—and this has to be an approximation since there's nothing to weigh it on 6,588,000,000,000,000,000,000 tons.
3. It would take Jackie a thousand years—isn't that amazing? You have to be amazed by how much money that means some people have.
4. Approximately 150 and 1,000. And exactly 153 and 1,122.
5. Approximately 25. And exactly 25, too. They are 2, 3, 5, 7, 11, 13, 17, 19, 23, 29, 31, 37, 41, 43, 47, 53, 59, 61, 67, 71, 73, 79, 83, 89, 97.
6. About 5 pieces per person, and exactly $6\frac{2}{3}$ pieces per person.
7. Approximately 30, because it's somewhere between 20 and 40. And exactly $26\frac{2}{3}$ minutes per errand.
8. Approximately 100 books, and measuring by a standard book and a standard suitcase, exactly 91 books. Do you need to stuff a suitcase full of books? Read up on the geometry chapter about volume and find out how.

CHAPTER 2

1. For medium-sized Granny Smith, it's about $\frac{1}{4}$. Three bites of a little Macintosh apple is about one half.
2. The number of people in the world is about 6 billion, and $\frac{1}{8}$ of that is about 750 million. Probably not able to fit on your block, unless you live somewhere with extremely big streets.

3. These calculations were done with a Hershey Bar and an old Kansas record, so results may vary. But if you guessed about 12 candy bars per record, you were right. And if you guessed 2 candy bars per compact disc, you were right again.
4. Well this is going to depend on the room you are in, right? To fill up the room I'm in, it would take 864,000 marshmallows of your standard campfire size. And I am in a very small room, 6′ by 10′ and 8′ high.
5. It would take approximately 288,000 golf balls to fill up the standard elevator.
6. There are exactly 100,000,000 pennies in $1 million.
7. The brain is approximately 2% of the average person's body weight.

CHAPTER 3

1. The ratio of men to women in the U.S. Senate in 1992 was 98 to 2. Yikes.
2. The ratio of men to women in the U.S. Senate in January 1993 was 94 to 6. Yikes again.

CHAPTER 4

1. You would have exactly $214,783,648. Exponential growth is pretty hot stuff.

CHAPTER 5

1. The average lead pencil could draw a line 35 miles long.
2. The average person consumes a ton, or 2,000 pounds of food and drink per year.
3. For the average person, the ratio of the measurement from fingertip to fingertip to height is exactly 1 to 1. This means when you stretch out your arms you are as wide across as you are tall.
4. Approximately 11,000 words are spoken in 1 hour on the radio.

CHAPTER 6

1. The heart beats 100,000 times every 24 hours.
2. If you guessed anywhere between 40° and 50°, terrific. The angle measures exactly 45°.
3. The perimeter is exactly 20, since all of the sides are equal.
4. There are 119 grooves around the circumference of a quarter.

CHAPTER 7

1. There are 2,598,960 possible different five-card poker hands.
2. This is frightening. There are 170,000,000,000,000,000,000,000,000 ways to make the first ten moves in a game of chess.
3. The average height for women is 5′ 4″, the average height for men is 5′ 10″.

INDEX

About the Author

Marcia Lerner graduated from Brown University in 1986. She has been teaching and writing for The Princeton Review since 1988. She lives in Brooklyn, N.Y.